Livestock/Deadstock

In the series **Animals, Culture, and Society,**
edited by Arnold Arluke and Clinton R. Sanders

Livestock/ Deadstock

Working with Farm Animals from Birth to Slaughter

Rhoda M. Wilkie

TEMPLE UNIVERSITY PRESS
Philadelphia

TEMPLE UNIVERSITY PRESS
Philadelphia, Pennsylvania 19122
www.temple.edu/tempress

Library of Congress Cataloging-in-Publication Data

Wilkie, Rhoda.
 Livestock/deadstock : working with farm animals from birth to slaughter /
Rhoda M. Wilkie.
     p. cm.
 Includes bibliographical references and index.
 ISBN 978-1-59213-648-3 (hardcover : alk. paper) -- ISBN 978-1-59213-649-0
(pbk. : alk. paper) 1. Livestock. 2. Food animals. 3. Human-animal relationships.
I. Title.
 SF84.3.W55 2010
 636.001'9--dc22

                                                          2009048498

Printed in the United States of America

2 4 6 8 9 7 5 3 1

To Colin and William

Contents

Acknowledgments

I would like to take this opportunity to thank each and every contact who assisted me during my period of fieldwork and interviews. Without their time and contributions, this book would not have been possible. I also thank the Carnegie Trust for awarding a Carnegie Scholarship. I am grateful to many colleagues at the University of Aberdeen for their interest and support, especially Bernadette Hayes and Debra Gimlin; I am particularly indebted to Steve Bruce for taking the time to read and comment on draft chapters. I thank Janet Francendese of Temple University Press and Clinton Sanders for their insightful feedback on an earlier version of the manuscript. For all of this support, I am most grateful. I also thank family and friends for being there when I needed them, especially Alistair and Fiona Wilkie, Kim Wilkie, Lesley Tidmarsh, Alastair Matthewson, Cath Pilley, Sandra Nicoll, and John and Triinu Humphrey. Finally, I express my sincerest gratitude to Colin Carroll. I appreciate his patience and words of encouragement much more than he will ever know.

Livestock/Deadstock

1

Food Animals

More Than a "Walking Larder"?

> As the public became aware of the horrific suffering of calves
> sent overseas to be incarcerated in veal crates, campaigns sprung
> up around the country. . . . But it was at Coventry airport that
> committed animal rights campaigner, Jill Phipps, died as she
> tried to stop a lorry carrying calves from entering the airport. Jill
> ran in front of the truck expecting it to stop but the convoy kept
> moving and she was crushed beneath the wheels, dying instantly.
> (Animal Aid 2006, 3)

During the mid-1990s, people in southern and southeastern England
took to the streets to protest against the export of live animals
through their towns. At the peak of the protests, more than 2,000
people congregated every night in the Essex port of Brightlingsea to
impede trucks taking veal calves to Europe (Brown 2006, 1). The antics
of protesters and police attracted extensive local, national, and international media coverage, making this one of the United Kingdom's highest
profile animal-related campaigns in recent years. That many campaigners (such as pensioners and mothers with young children) had never
protested before challenged the media stereotype of the typical animal-
rights campaigner. Slogans such as "You don't have to stop eating meat to
care—ban live exports" and "On the hook not the hoof" indicate that the
campaign was not overtly framed as a vegetarian issue and may partly
explain the broad spectrum of public support this issue successfully
mobilized. One resident observed, "There is no escape from the daily
convoys with the sight, sound and smell of the animals" (McLeod 1998,
46).[1] Farm animals became an urban spectacle, and the trade itself, along
with related animal-productive contexts and practices, became subjected
to increasing public scrutiny.

These protests sparked my academic interest in human–livestock relations. In the summer of 1995, I was doing research for my undergraduate
dissertation in sociology and was especially interested in protest groups

and how they were organized. The campaign to ban live exports seemed an obvious and appropriate comparative site for my project. I traveled to Brightlingsea to establish contacts, and spend time, with those involved in the protests at that time. Having negotiated access, I became curious about why women were so prevalent in this campaign and in issues pertaining to animal rights. Such questions were to be the focus of my doctoral thesis. However, it subsequently dawned on me that many campaigners had no pre-existing relationship to the animals on whose behalf they protested. I then began to wonder how agricultural workers who do have direct contact with livestock make sense of their interactions with them. This insight signaled a change in research focus that ultimately took me from protest to production (Wilkie 2006, 3).

This book opens up an under-explored and little understood aspect of contemporary life: men's and women's relationships with farm animals. Many people have firsthand experience with companion animals such as dogs and cats, but few have encountered animals reared and killed for food. In 1900, more than "half the population made a living producing food, [but] this has dramatically changed. Today barely 1.7 percent of the population works in production agriculture, with perhaps half of that group or less in animal agriculture" (Rollin 2004, 7). As consumers, we no longer need to produce the food we eat because others are paid to do it for us. Instead, we go to supermarkets and shops to buy a wide range of animal-derived foodstuffs, such as milk, eggs, cheese, and meat. As the link between production and consumption, and between producers and consumers, lessens, food animals, and those who work with them, have increasingly slipped from public view and consciousness. The more divorced we are from animal-productive processes, the easier it becomes to dissociate the animals we eat from the meat on our plates. As Carol Adams (2000, 14) notes, "Behind every meal of meat is an absence: the death of the animal whose place the meat takes. The 'absent referent' is that which separates the meat eater from the animal and the animal from the end product."

This book is not about the rights and wrongs of rearing, killing, and eating food animals. Such ethical debates are profoundly important, but related discussions have tended to be framed in fairly abstract terms that fail to take adequate account of people's hands-on experiences of working with livestock. What has been missing from the wider picture is a more nuanced appreciation of how those who breed, rear, show, fatten, market, medically treat, and slaughter livestock perceive and make sense of their interactions with the animals that constitute the center of their everyday working lives. Gaining insight into the nature of commercial and hobby human–livestock relations from the perspectives of small-, medium-, and large-scale producers, auctioneers, livestock auction workers, vets, and slaughterers is a noteworthy piece of the jigsaw puzzle, a piece that needs far more scholarly attention. This book begins to explore this interspecies blind spot from a sociologically informed perspective. I suggest that one way to understand the various attitudes, feelings, and behaviors of those who work with livestock (in this case, mainly cattle, and to a lesser extent, sheep) is to consider the location of both people and animals in the commercial and non-commercial productive division of labor—that is, from birth to slaughter. Commercial and hobby contexts give agricultural workers (mainly male) and hobby farmers (mainly female) varying opportunities to interact with, and disengage from, the different species and breeds of livestock with which they work. This provides a more nuanced insight into the multifaceted, gendered, and contradictory nature of human–animal productive roles in modern industrialized societies. Livestock animals are not all the same. Neither are the people who work with them; nor are the productive contexts in which humans and animals work.

Exploring how different groups of agricultural workers think, feel, and relate to food animals also provides an additional perspective on people's dealings with domesticated animals. For example, domesticated animals are usually categorized as either livestock or pets. However, this clear-cut dichotomy is messier in practice because many of my contacts perceived some of the livestock they worked with as pets, friends, or even work colleagues. This highlights three interesting points: Livestock can be more than just "walking larders" (Clutton-Brock 1989); the animals' status as commodities is unstable; and instrumental and socio-affective attitudes can and do coexist, albeit to varying degrees, in these different productive settings. Underestimating the range of roles people can ascribe to livestock misrepresents and oversimplifies the potentially complex and dynamic nature of human–livestock interactions. These findings begin to challenge commonly held assumptions that productive animals are nothing more than mere commodities and that agricultural workers are typically uncaring toward such animals. Unlike hobby farming, commercial systems of animal production do depend to a great extent on instrumental attitudes of those working in this sector. But such attitudes can be disrupted and temporarily suspended when some animals come to be seen as individuated beings. I suggest that much can be gained from exploring the multifaceted and ambiguous nature of people's practical relations with livestock because it provides an opportunity to shed additional light on, and offer fresh insights into, longstanding dilemmas and ethical debates about the production and slaughter of food animals in modern industrialized societies.

For instance, the way food is produced in America is generating much unease among consumers (e.g., Pollan 2006). The production of livestock is no exception, especially that of animals that have been intensively produced in confined animal-feeding operations (CAFOs).[2] According to a report by the Union of Concerned Scientists, "CAFOs are distinguished from other ways of raising livestock by their size, high-density confinement of livestock, and grain-based diet, which requires bringing feed to the animals rather than allowing animals to graze or otherwise seek their food" (Gurian-Sherman 2008, 13). Over a period of a few decades, CAFOs became the main method of production in the United States. Although they account for about 5 percent of all animal operations, they produce more than 50 percent of all food animals. One of the key issues the report highlights is the "hidden costs" that have been factored out, or "externalized," to society, which understate the real costs associated with this type and scale of animal production (Gurian-Sherman 2008, 3–5). Such costs can take many forms: taxpayers footing the bill for feed grain subsidies; streams and rural water supplies being contaminated with animal manure; respiratory ailments being triggered by airborne ammonia mixing with other sources of air pollutants; and antibiotic-resistant pathogens such as methicillin-resistant *Staphylococcus aureus* (MRSA) being created by excessive use of antibiotics.[3] In addition, consumers deliberate the welfare and conditions of intensively produced farm animals and discuss "what constitutes a decent life for animals and what kind of life we owe the animals in our care" (Pew Commission on Industrial Farm Animal Production 2008, 13).

The wide-ranging fallout from CAFOs has been a catalyst for present-day debates about how livestock are produced in America. In the United Kingdom, it was the emergence of "Mad Cow Disease," or bovine spongiform encephalopathy (BSE), that focused the public's attention on how food animals were raised and disposed. First identified in Britain in 1986, BSE reached epidemic proportions by the mid-1990s. "By 2000 it had killed over 200,000 cattle and caused the slaughter of almost four million exposed animals" (Kahrs 2004, 60).[4] Scientists investigating this new cattle disease discovered that the brains of infected cattle showed neuropathological changes comparable to those in sheep with scrapie, a long-known spongiform encephalopathy (Smith and Bradley 2003, 188). Three of the main clinical signs of the disease were "nervousness, heightened reactivity to external stimuli, and difficult movement, particularly of the hind limbs" (Pattison 1998, 390). Once BSE manifests in cattle, the course of the disease can vary. Even so, as the disease progresses, it gets worse; it is incurable, and thus infected animals are slaughtered (Department for Environment, Food, and Rural Affairs 2007a, 3). Researchers think the BSE agent could be "a self-replicating protein, referred to as a prion" (World Health Organization 2002, 1),[5] and the most likely mechanism for spreading the disease is meat-and-bone meal (MBM), a high-protein supplement feed made by rendering inedible bodily waste materials of different species of livestock (Smith and Bradley 2003, 186).[6]

During the 1970s and 1980s, fallen stock (animals that died on the farm) was rendered as animal feed. As no evidence at that time indicated that scrapie could be transmitted to other species of food animal, sheep carcasses infected with scrapie became part of MBM, and for the first time, cattle were exposed to bodies of ruminants.[7] The temperatures during the rendering process were set below those recommended in European guidelines, and this may have contributed to the "survival of the scrapie agent (or a mutation thereof) in feed and its dissemination among cattle" (Kahrs 2004, 60). Researchers surmised that once the BSE agent entered the rendering process, it was transmitted to other cattle when they ingested infected MBM. The infected waste tissues of slaughtered bovines would in turn be added to the rendering process, which facilitated the ongoing spread of the disease via MBM. This animal-feed death loop led to a ban on ruminant feed in 1988. However, the ban was limited to ruminant species; proteins derived from rendered sheep or cattle could still be fed to non-ruminant species such as pigs and poultry.[8]

In 1996, eight years after the ban, the first case of new variant Creutzfeldt-Jakob disease was identified in Britain.[9] The appearance of this incapacitating brain disease in a human being indicated that BSE could be transmitted from animals to people. The U.K. government responded by imposing a complete ban "on feeding mammalian protein to any farmed animals." All other European Union (EU) countries followed suit with a similar ban in 2001 (Smith and Bradley 2003, 189–191). Because this was a new disease, experts and officials were not sure how best to manage the unfolding crisis, which led to contradictory messages being passed on to the public. Consumers responded by drastically reducing their intake of meat to the extent that sales plummeted by about 40 percent. There was also a worldwide ban on the exporting of all live cattle and cattle-derived products from the United Kingdom to other countries (Franklin 1999, 169). To further minimize the threat to human health, the U.K. government decided that all cattle over thirty months old could no longer enter the food chain.[10] According to Sir John Pattison (1998, 392), "This added an extra margin of safety because cattle can be reasonably accurately aged by their dentition at 30 months and because BSE is relatively rare under the age of 30 months." This strategy required such animals to be incinerated after slaughter; in effect, their entire bodies were treated as specified risk material (SRM). In some ways, this was an extension of an existing intervention to safeguard consumers from cattle organs most likely to be infected by BSE: the brain, spinal cord, tonsils, thymus, spleen, and intestines. In 1989, the United Kingdom introduced a specified ban on bovine offal.[11] In practice, this meant that high-risk bodily organs, believed to account for "99% of the infectivity in a BSE-infected bovine," had to be stained blue in abattoirs to signify its SRM status (Smith and Bradley 2003, 194).[12] In other words, such material was deemed unfit for human consumption.

BSE and other food-related scares, such as those involving *Salmonella, E. coli,* and foot-and-mouth disease, have prompted more and more consumers to question the safety of "food with a face" and its production (Williams 2004, 46). Many people were astonished and appalled to discover that cattle, herbivore animals, were being fed the bodies of reprocessed ruminants. This does not sit comfortably with idealized images of cows methodically grazing rich pastures on a sunny afternoon. An important consequence of these scares was that it brought livestock and related productive practices back into public consciousness: "They emphasized to the consumer, the connections between animals and meat, and underlined the processes of animal-into-meat" (Franklin 1999, 164). This theme of connecting and reconnecting is reflected in one of the main recommendations put forward in *Farming and Food: A Sustainable Future,* also known as the Curry Report: "The key objective of public policy should be to reconnect our food and farming industry: to reconnect farming with its market and the rest of the food chain; to reconnect the food chain and the countryside; and to reconnect consumers with what they eat and how it is produced" (Policy Commission on the Future of Farming and Food 2002, 6).

But reconnecting with the animate basis of our food can be far from straightforward, especially if it requires the death of healthy animals—which, of course, meat does. As I show in the chapters that follow, consumers and policymakers are not alone in this animal-productive-consumption dilemma. Many of those involved in producing, marketing, medically treating, and slaughtering livestock grapple with it, too.

Given this background, issues pertaining to the production, welfare, and slaughter of food animals have been ascending political and public agendas in Britain, as well as in the EU and in other industrialized countries. As I show in Chapter 6, such concerns have a long history. Even so, Ruth Harrison's *Animal Machines* (1964) is one of the first seminal critiques of contemporary animal-production methods in the United Kingdom. In her opening pages, she observes with concern the emergence of the "factory farm":

> Farm animals are being taken off the fields and the old lichen covered barns are being replaced by gawky, industrial type buildings into which the animals are put, immobilised through density of stocking and often automatically fed and watered. Mechanical cleaning reduces still further the time the stockman has to spend with them, and the sense of unity with his stock which characterises

the traditional farmer is condemned as being uneconomic and sentimental. Life in the factory farm revolves entirely round profits, and animals are assessed purely for their ability to convert food into flesh, or "saleable products." (Harrison 1964, 1)

As I discuss in Chapter 2, the shift from animal husbandry to animal industry breached the "social contract" thought to characterize traditional farmer–livestock relations (Rollin 1995). As a consequence, ethical and welfare dimensions of animal productive contexts that once were pretty much taken for granted have become subjected to increasing public, scientific, and political scrutiny. For example, critical appraisals of livestock farming, including high-profile animal-welfare campaigns, have tended to focus attention on the most intensively produced farm species: "veal crates for calves, stall and tether-cages for pregnant pigs, and battery cages for laying hens" (Druce and Lymbery 2006, 123; see also Singer 1995). As such groups often harness to good effect shocking exposé media coverage of such issues (Baker 1993), this has contributed to activating, mobilizing, and registering a broad range of public concerns about some of the more questionable practices occurring within the industry. At one end of this diverse spectrum are consumer groups and farm-animal-welfare organizations that lobby governments and producers to improve how livestock are treated in such settings. At the other end of the animal-welfare-rights continuum are those who are critical of simply tinkering with the existing system; by implication, this continues to endorse the use of animals for food and legitimates their legal status as property (Francione 2006). Advocates of a more absolutist standpoint believe it is irrelevant how humanely livestock are reared and slaughtered because it is morally unacceptable to kill them to put on your plate. By promoting meat-free diets, such as vegetarianism and veganism, proponents of this perspective challenge dominant ethical norms and species assumptions that underpin and reproduce the presupposed practice of killing and eating livestock (Nibert 2002; Regan 2004). However, "there is a significant chasm between the theory of animal rights and the social phenomenon that we call the 'animal rights movement'" (Francione 1998, 45). Proponents of "new welfarism," a fusion of these positions, believe that securing incremental changes that improve the welfare of animals in the interim are necessary steps toward realizing more radical animal rights goals in the long-term. Even though such advocates might still use the language of animal rights they seem to adopt a more pragmatic approach to bringing about more deep-seated changes.

Supporters of such groups have been the industry's keenest observers and its most vociferous critics. Researchers note that many have little experience with the farm animals on whose behalf they protest (Webster 2005), are more familiar with pets (Jamison and Lunch 1992), are predominantly female (e.g., Herzog et al. 1991; Shapiro 1994), and are from urban settings (Kendall et al. 2006; Serpell 2004). Research also has found that "consumers do not trust governments, the EU, or the food industry as sources of information on standards of animal welfare. They are much more willing to trust consumer and animal welfare organisations" (European Commission 2002, 18). Public skepticism has clearly contributed to opening up a critical space and influential platform for such non-official groupings. Their contribution to the public debate has been invaluable, and they have undoubtedly ensured that "the status of commodified domestic animals such as cattle, sheep, pigs and chickens, once excluded from spheres of moral concern and legal protections, is being re-evaluated" (Emel and Wolch 1998, 14).

However, animal advocates, and related organizations, also promote their critical perspectives and narratives of contemporary livestock production in publicly available reports, books, and articles. The increasing prevalence of such information "has created a 'New Perception' of animal agriculture by depicting commercial animal production as 1) detrimental to animal welfare, 2) controlled by corporate interests, 3) motivated by profit rather than by traditional animal care values, 4) causing increased world hunger, 5) producing unhealthy food, and 6) harming the environment" (Fraser 2001, 634).

Scholarly members of the academic community, such as scientists and ethicists, read these resources, too. However, if academics uncritically utilize or overly rely on advocacy literature in their analyses, they run the risk of perpetuating "misleading, polarized, or simplistic accounts of animal agriculture" (Fraser 2001, 634). As David Fraser notes, the New Perception has brought to the fore some searching questions about the nature of modern-day methods of livestock production, but the quality of the debate so far has been hampered by partisan positions that undermine a more sophisticated and nuanced discourse on such issues. Similarly, animal scientists conducting research for users within the agricultural sector have tended to produce research that more readily addresses the interests of those within the industry as opposed to those outside it.

Furthermore, one-dimensional accounts are more likely to depict intensive methods of production as being bad and workers in such systems as a relatively uncaring homogenous cohort who treats all species of livestock as an undifferentiated commodity that is exploited for profit. If this is the case, it implies that "the commercialization and industrialization of the livestock industry has created a class of animal producers wholly insensitive to animal needs" (Thompson 2004, 149). However, some researchers note that small family-owned farms are less harmful to their animals compared with large-scale producers (DeGrazia 2003, 182). Although the scale of industrialized production would intensify any negative effects on a greater number of animals, it is possible that some intensive units may adhere more stringently to animal-welfare legislation than some family-run farms. Like any other well-managed organization, such operations would be more likely to encourage staff to comply with legal requirements and industry best practices and discourage corner-cutting practices. Similarly, small- to medium-scale producers may be more positively perceived because traditional farmers are believed to practice a higher level of animal-husbandry skills than their more hyper-commercial counterparts, but this cannot always be assumed. Drawing such clear-cut distinctions between factory farms and family farms also seems less straightforward in practice. Globally, only a minority of CAFOs are corporately owned. By and large, "Owner-operators of CAFOs farm under contract to firms that integrate the various links of the food chain" (Thompson 2004, 149). These firms tend to assign contracts to family-owned and family-run farms. In contrast, "Integrators generally own a fairly small component of the total chain, and it tends to be the slaughtering and processing facility, not the 'factory farm'" (Thompson 2004, 149). Thus, it seems that many factory farms are likely to be operated and overseen by owners of family farms.

Livestock-production systems vary widely, too, especially with respect to the extent to which animals are intensively produced. Some species of livestock are biologically better suited to intensive production than others. According to the Food and Agriculture Organization, "The major expansion in industrial systems has been in the production of pigs and poultry since they have short reproductive cycles and are more efficient than ruminants in converting feed concentrates (cereals) into meat" (Bruinsma 2003, 166). Danielle Nierenberg (2005, 11–12) notes that industrial approaches account for "74 percent of the world's poultry products, 50 percent of all pork, 43 percent of beef, and 68 percent of eggs." The more intensively farmed species (i.e., poultry, pigs, veal calves, and dairy animals) also have the highest public profile. Sheep and beef cattle tend to attract less forthright public condemnation; perhaps the production of these species conjures up "a more natural image of animals in fields eating grass" (Webster, 1994, 128). For instance, in the United Kingdom, spring-born calves in suckler beef systems will suck milk from their mothers (or dams) while out to pasture over the summer.[13] These calves are usually weaned from their mothers at six to ten months old and housed inside during the winter. The young animals are then finished off, or fattened up for slaughter, on grass over the coming summer months.[14] Perhaps the public perceives animal systems with an extensive component in a more positive light—that is, as more natural and animal welfare-friendly. However, extensive systems can be problematic, too. For example, research into extensive beef cattle systems in northern Australia has found that climatic conditions can disrupt animal welfare "through thermal stress and food shortages" (Petherick 2005, 212). In addition, free-range cattle are minimally handled and widely dispersed, conditions that hinder producers from identifying and treating animals that are unwell. Even low-input, extensive productive systems that impose minimal restrictions to the animal's "natural" functioning can throw up animal welfare-related concerns (see Rollin 2004).

At the other end of the production process is the highly intensive and much criticized feedlot approach to finishing cattle in the United States. The High Plains (or "Beef Belt") epitomizes the hothouse of beef-cattle production in regions such as southwestern Kansas, Texas, and Oklahoma. Some of the large operations in Kansas have the capacity to feed and finish up to 112,000 head of cattle for slaughter (Harrington and Lu 2002, 275). The bulk of feedlots are to be found in states with abundant grain supplies (Animal and Plant Health Inspection Service 2007, 76), and according to statistics from the Food and Agriculture Organization, 70 percent of all corn harvested in America is used to feed livestock because such grains promote rapid weight gain (Nierenberg 2005, 23). For example, "Beef calves can grow from 36 kilograms to 544 kilograms in just 14 months on a diet of corn, soybeans, antibiotics, and hormones" (Nierenberg 2005, 23). But such highly concentrated finishing regimes can induce a range of well-recognized animal-related problems—for example, digestive ailments, including liver abscesses, laminitis (i.e., inflammation of the hooves that can cause acute or chronic lameness), heat-induced stress from prolonged exposure to intense sunlight, and behavioral abnormalities such as tongue rolling, bar sucking, and aggression (Webster 2005).

As mentioned, the production of livestock generates much concern about, and puts pressure on, key environmental resources: "Livestock have a substantial impact on the world's water, land and biodiversity resources and contribute significantly to climate change" (Food and Agriculture Organization 2006, 4; see also Pew Commission on Industrial Farm Animal Production 2008). Researchers have also estimated that this sector is accountable for 18 percent of anthropogenic greenhouse gas emissions (Food and Agriculture Organization 2006, 112). On top of this, the "global production of meat is projected to more than double from 229 million tonnes in 1999/01 to 465 million tonnes in 2050, and that of milk to grow from 580 to 1043 million tonnes" (Food and Agriculture Organization 2006, xx). Growing demand for food-animal products is being driven by a number of factors. First, the current world population of 6.5 billion is anticipated to increase to 9.1 billion by the year 2050 and could reach 9.5 billion by 2070 (Food and Agriculture Organization 2006, 7). Second, as countries such as China become richer and more urban, they develop an affluent stratum of people who have money to spend on a wide range of consumer items, including meat and dairy products.[15] Third, the preference for animal-derived produce is characteristic of a dietary trend called "nutrition transition": As people in transitional countries become wealthier, they eat fewer locally available grains, roots, vegetables, and fruit and consume "more pre-processed food, more foods of animal origin, more added sugar and fat, and often more alcohol" (Food and Agriculture Organization 2006, 10). The increased global demand for food-animal products has produced a "Livestock Revolution" and is forecast to "stretch the capacity of existing production and distribution systems and exacerbate environmental and public health problems" (Delgado et al. 1999, 1).

This brief overview indicates that, in Western agricultural contexts such as Europe, Australia, and the United States, there is a continuum of highly industrialized to less industrialized to non-industrialized livestock-production approaches along which small- to medium-scale farming units can coexist—admittedly rather precariously at times—alongside larger and more intensive producers. Each farm-animal species can also be produced to varying degrees of intensiveness and extensiveness depending on which management system is adopted. John Webster (2005, 98–99) suggests that each type of productive context has different strengths and weaknesses in terms of how each system might affect the well-being of the animals produced therein.[16] Such an analysis encourages a more even-handed appraisal and reminds us that even seemingly natural methods can be problematic, too. It also mitigates the possibility of perpetuating a somewhat one-dimensional understanding of animal-production methods summed up as "intensive bad, extensive good" (Webster 2005, 98–99).

Equally, one-dimensional depictions of farmers, and other livestock-related personnel, fail to take adequate account of a range of individual (e.g., age, gender, and experience in handling livestock) and institutional factors (e.g., commercial or hobby production, position in the division of labor, scale of animal production) that could influence the different types of attitudes, feelings, and behaviors they might have toward the different species of animals with which they work. As I show over the course of this book, different stages of the production process provide a range of opportunities and constraints on the extent to which animal handlers actively engage with, or dissociate from, the animals they manage. As David Fraser (2001, 638) notes, there has been a tendency "to treat animal agriculture as an aggregate and draw conclusions that are unwarranted because they are unduly general." Given that most of us have little, if any, experience in livestock farming, we are reliant on second-hand sources for our information. Currently, a lot of this information seems to come from animal-welfare and animal-rights organizations that tend to focus on and critique factory-farming approaches. Although such contributions have been very significant, they do not represent all of the productive contexts in which different species of animals are produced or the full range of human–livestock relations that might occur within these settings. We need to be alert to the possibility that our contemporary understanding of food-animal-productive contexts is being perceived primarily through a factory-farming lens.

The largely critical nature of this coverage has left many industry workers feeling rather defensive and circumspect in their dealings with outsiders. At the same time that members of the general public and official bodies must be made aware of the darker sides of animal production, including bad practice, agricultural workers in general must not be routinely cast as villains. Coverage of the industry gives little attention to examples of proficient animal handling and fails to report that many workers carry out their duties competently and others go beyond this by practicing "good stockmanship" (discussed in the next chapter). Perhaps the industry has responded to unfavorable exposure by promoting a rather oversimplified and more positive image of animal production than is justified by actual conditions (Fraser 2001). For instance, the use of talking animals in industry advertisements, such as the California Milk Advisory Board's "Happy Cows" campaign, might be an effective way to promote sales of cheese, and of engaging its target audience's attention, but this medium has generated much criticism, too, because it does not accurately represent the reality of the productive contexts in which animals are farmed (Glenn 2004).[17] The livestock sector is not oblivious to or unaffected by wider changes in society or shifting moral expectations of its customer base. By conforming to industry ideals and legal regulation, it treats livestock in a manner that promotes their welfare and minimizes unnecessary suffering; this standard not only enhances animal productivity but also boosts financial returns and improves the public image of the industry (e.g., English et al. 1992; Hemsworth and Coleman 1998).[18] The industry wants the best of both worlds. But, of course, industry opponents argue that any degree of "unnecessary suffering" is unjustifiable because "all of our animal use can be justified *only* by habit, convention, amusement, convenience, or pleasure. . . . Most of the suffering that we impose on animals is completely unnecessary" (Francione 2006, 79). What this clearly shows is just how polarized and contested contemporary food-animal production has become.[19]

In all, it would seem that our current knowledge of human–food animal relations is somewhat partial and partisan. As we have reached something of an impasse, this is perhaps an opportune time to take stock of people's relationships with livestock. What has been missing from the debate is an appreciation of how those involved in the daily tasks of producing and slaughtering make sense of their interactions with and experiences of the animals they work with as part of their everyday lives. This in essence was the purpose of my research and leads me to the nature of my study. Here, I should note that most of the commercial producers in my study worked with beef cattle. Others, such as auctioneers, mart workers, and veterinarians, worked with different species of livestock, especially cattle and sheep and, to a lesser extent, pigs. The reason for this difference is that I gained access to the commercial sector via one of the most modern and largest livestock auction markets in Europe. Since intensively produced pigs and poultry tend to bypass the live auction market, I mostly came into contact with people who worked with ruminant species.[20] However, my contacts in hobby farming were more likely to work with sheep, goats, pigs, poultry, and rabbits. My research therefore provides insight into animal-productive contexts that are less associated with high-profile factory-farmed species (such as pigs, poultry, dairy, and veal calves) and methods.

I conducted most of my research in northeastern Scotland, which has a worldwide and long-standing reputation for the production of beef cattle (especially Aberdeen Angus) and is thus a pertinent site for such research (e.g., Carter 1979; M'Combie 1875; Perren 1978; Trow-Smith 1959). I chose an ethnographic approach characterized by "the ethnographer participating overtly or covertly, in people's daily lives for an extended period of time, watching what happens, listening to what is said, asking questions—in fact, collecting whatever data are available to throw light on the issues that are the focus of the research" (Hammersley and Atkinson 1995, 1). From August 1998 to August 1999, I combined a four-month period of overt participation observation with forty-six in-depth but fairly unstructured interviews with fifty-three people.[21] At that time, the commercial sector was still coming to terms with the fallout and bureaucratic implications of the BSE crisis, and the strength of the pound and poor weather conditions contributed to financial difficulties within the industry (Slee 1998). I was therefore acutely aware that a predominantly, but not exclusively, male-dominated workforce whose livelihood was under threat might be reticent toward a female researcher who was a relative outsider. Although I was born into a farming family, we moved to the city during my early primary-school years. People in the field frequently asked whether I came from a farming background; having a father who farmed in the area and a few extended members of the family who still do so helped my respondents locate me in their world. My ability to speak and understand Doric, the local rural dialect, reinforced my local country roots. However, as I noted in my field notes, "My identity is confused in that I come from the country, still have a country accent, but have lived in the town for most of my life. My parents have lived their lives in the country—I feel a connection with these people yet there are times I feel so distant. . . . I have all the right credentials but I have little lived experience of living and working in the country."

In some ways, I was an honorary member of the farming community through birth and family connections, but such membership was in one respect misleading. It was assumed that I ate meat. In my late teens I started to remove meat from my diet, and I have become increasingly vegetarian. In later years, I adopted a vegan diet. As Nick Fiddes (1991, 111) suggests, "True vegans aim totally to avoid all animal produce—including dairy products, wool, leather, even honey, and sometimes extending to such items as batteries or beer which are reputed to use animal products in manufacture." Based on this definition I am vegan by diet only, not by lifestyle. If people enquired during my fieldwork whether I was vegetarian, I agreed to this label for the following reason: I am not strictly vegan, but vegans by virtue of their diet are vegetarian. I was the embodiment of ambiguity: I passed as "one of them," but at the same time I was seen by some in this context as being "one of them." However, as Geoffrey Pearson (1993, xviii) suggests, "The ethnographer does not have to be a competent burglar, or prostitute, or policeman, or miner in order to deliver competent ethnographies of work, life, and crime. . . . It is an old adage of social research that you do not need to be Caesar in order to understand Caesar; indeed, it might even be a handicap."

Given all this, I thought it prudent to carry out a period of fieldwork to increase my general understanding and knowledge of the livestock-production process. As the livestock auction is the public face of the industry, and its social hub, it was an obvious entry point into the sector. Having gained permission from management to conduct my research in this setting, I worried that my presence might raise suspicions about my purpose among members of the workforce (May 2001, 157). Indeed, management's concerns for my safety and restriction of my research role in the mart to a primarily observational one probably gave workers further reason to be wary of me early in my fieldwork. The mart has the capacity to process 3,000 cattle and 14,000 sheep. Given the unpredictable nature of working with livestock, especially cattle, management permitted me to spend time with mart workers backstage in the market as long as I remained vigilant when the animals were being moved. I soon identified the safest observational points and quickly realized when I was standing in the wrong place. During the early stages of my fieldwork, many workers seemed to assume I was monitoring animal welfare and was thus a tool of management. Over time, mart workers more readily acknowledged me to the point at which I was able to engage them in impromptu conversations, which helped to dispel their suspicions. Such workers are relatively unaccustomed to having a student spending protracted periods of time in their work domain and expressing an interest in their work. Handling livestock is perceived as a low-status job, so why would anyone from an academic background want to hang around in a cold, smelly, noisy, and potentially dangerous setting?

Having become familiarized with and better informed about the various roles of those involved in handling livestock from birth to slaughter, I was ready to carry out my interviews. I talked to farmers and stockmen, people working at the mart, hobby farmers, veterinary staff, and a few abattoir workers.[22] Field contacts and interviewees also recommended additional people I could interview. This snowball technique is useful, but it bypasses people who are not connected with my contacts' specific social networks. If my contacts limited their recommendations to people they thought would be most open to being involved in the study, my findings might be skewed to those who could be described as "good stockmen" or "good farmers." As there has been so little sociological research into human–livestock relations in commercial and hobby contexts, even firsthand accounts from good stockmen offer a useful insight into this neglected area. My approach to interviewing was fairly flexible, as I wanted the opportunity to explore any relevant issues raised by my contacts during the interview. Moreover, interviewing people from diverse working contexts required the flexibility to ask questions that were appropriate to each of their particular work environments. Notwithstanding, I tried to explore some core themes with all interviewees: personal background, experience working with livestock, stages associated with producing livestock, how they made sense of their interactions with the animals they worked with, and the slaughter of livestock (see Wilkie 2002). The interviews, which lasted between forty-five minutes and three hours, were audiotaped and transcribed. I analyzed data in an ongoing iterative process as I carried out my fieldwork and interviews. Thus, I adopted a grounded analysis that enabled me to explore, cross-reference, and test the accuracy of my observations and emerging ideas with research contacts and interviewees.

Supplementing our existing knowledge of human–livestock interactions with emerging empirical findings, from those located within the commercial livestock sector and from those marginal to it, will help to broaden and deepen our current understanding of this rather neglected interspecies connection (see e.g., Lovenheim 2002; Tovey 2003). In the first half of the book, I trace the rise of this ancient relationship by outlining some of the main perspectives and debates surrounding the process of animal domestication. This leads to a historical overview of some of the key socioeconomic and technical factors that contributed to the industrialization of cattle production and slaughter in the United States and Europe and the subsequent commercialization of human–livestock relations. I then explore the gendered nature of food-animal production and women's roles in livestock farming. The livestock auction market is not only the public and male face of the commercial sector; it is also where livestock are marketed and economically valued. I trace the rise of the modern auction system, and auctioneering, in Britain and some of the factors that go into discovering the price of livestock at different stages of the production process. The second half of the book explores the ambiguous status of livestock and how those at the "byre face" negotiate the extent to which they actively form emotional ties to or disconnect from the breeding, store and prime animals they work with.[23] This leads to a detailed discussion of how healthy and ill livestock become deadstock and illustrates how agricultural workers, hobby farmers, and slaughterers make sense of and manage this pivotal transitional stage.

2

Domestication to Industry

The Commercialization of Human–Livestock Relations

> Domestication is generally taken to be the historical mile-
> stone that marks the most profound and definitive transfor-
> mation in the relationship between humans and other species.
> Domestication is not only seen to symbolise the critical transition
> from simply taking from nature to actively controlling it, but is
> also generally taken to represent the move which most clearly
> distinguished humans from other animals. (Swabe 1999, 25)

The domestication of animals more than 10,000 years ago threw
up a number of intended and unintended biological, cultural,
and socioeconomic consequences that continue to resonate today
(Cassidy and Mullin 2007). In this chapter, I outline some of the main
perspectives on and debates about this significant interspecies legacy and
then trace the emergence of the Euro-American "animal industrial complex" by identifying some of the key sociohistorical, economic, and technical factors that contributed to the industrialization of cattle production
and slaughter in Western societies (Noske 1989; Rifkin 1992). As this
takes us to the hub of human–livestock relations, I explore the changing role and status of stockmanship in contemporary livestock farming,
especially, but not exclusively, in the United Kingdom.[1] In particular, I
consider how the taken-for-granted social contract that typically characterized the productive relationship between traditional farmers and their
animals became destabilized as methods of animal production increasingly intensified after 1945.

"Domestication" was absent from the English language prior to the
1500s (Anderson 1998), but endeavors to define its meaning commenced
more than a century ago (Bökönyi 1989). Until the late 1940s, archaeologists showed little interest in human-animal relations, as they saw it as the
domain of anthropologists and biologists (Clutton-Brock 1989). In spite
of this, most agree that about 10,000 years ago human societies shifted
from a nomadic, hunting-gathering lifestyle to a more settled way of life

that revolved around farming and domesticated animals (Clutton-Brock 1994; Mithen 1996). The Neolithic (literally, "New Stone Age") period refers to this epoch and is characterized by the "presence of an economy wholly or partly dependent on domesticated crops or animals, or both" (Balaam et al. 1977, 167). Prominent archaeologists in the 1950s such as V. Gordon Childe interpreted this economic transformation as a pivotal development that "laid the foundation for complex societies, and 'progress.' The key to this was the storage of animals and plants and hence their availability in ever-greater quantities" (Wilson 2007, 102). It was believed that the construction of permanent dwellings was a response to agricultural activity, but the accuracy and direction of this assumption has been subject to review. As factors previously neglected by archaeologists, such as space and time, have been taken into account, some now suggest that "architecture not only precedes agriculture but provides the technology necessary for agriculture to emerge" (Wilson 2007, 101).

The emergence of farming is also seen as a prerequisite for human civilization. As Raymond Williams notes, "To cultivate nature was to draw it into a moral order where it became 'civilised.' Indeed, it was the practice that signified culture itself, a term, which in its earliest European use, meant to cultivate or tend something—usually crops and animals" (quoted in Anderson 1998, 126). This Westernized reading of domestication is built on a central assumption that there are two distinctive realms of being that are reflected by the categories "culture" and "nature."[2] Of course, these domains are not just different; they are also hierarchically valued. Nature and everything associated with it is regarded as inferior to the more cultured and superior realm of humanity. To the degree that humans subdue wild and brutish constituents of the natural world, including their own instinctive natures, they transcend nature to become more cultured and refined beings. To do otherwise fueled ethnocentric ideas, because "the activity of domestication seems to have been taken as a fundamental criterion for ranking groups of people called 'races'" (Anderson 1997, 468). Thus, non-Western indigenous groups who lived nomadically by hunting and gathering were considered primitive and savage because they "stood at the beginning of social time, 'unevolved' through having themselves *remained* undomesticated" (Anderson 1997, 474).[3] In contrast, animals were destined to live out their instinct-oriented lives exclusively within the natural realm.[4] Domesticating plants and then animals enabled humans increasingly to extricate themselves from nature's chains.[5] It also fostered a more detached view of, and instrumental attitude toward, everything located within this domain. This greatly contributed to the objectification of and human mastery over nature's raw resources, and this included animals (Thomas 1983).

Although archaeologists and biologists generally agree about when domestication occurred, they often dispute its rationale, means, and initial geographical location (Clutton-Brock 1999, 10). Our understanding of early animal domestication mainly relies on the scientific examination and interpretation of such excavated remains as fossilized bones and teeth. Such remnants can provide some insight into "the occurrence and morphology of particular animals, the climate and the environment in which they lived and even the diseases or injuries from which they suffered" (Swabe 1999, 24). To further contextualize such information, researchers might supplement their analysis by drawing on contemporary anthropological studies of hunter-gatherers and ethological data. Based on such evidence, the wolf (*Canis lupus*) was probably the first species to be successfully domesticated, about 14,000 years ago in Central Europe. Such animals increasingly would have been used for hunting and guarding tasks (International Livestock Research Institute 2007, 25; Serpell 1998b, 137). By about 10,000–8,000 years ago, people had turned their attention to other animals. For example, during this period wild goats (*Capra aegagrus*) and sheep (*Ovis orientalis*) appear to have been the first ruminant species to undergo domestication in places such as the Fertile Crescent and Turkey, followed by pigs (*Sus scrofa*) in the Near East and East Asia,[6] and cattle (*Bos primigenius*) in the Fertile Crescent, Near East, and Africa. Finally, chickens (*Gallus domesticus*), which are descendants of the wild red jungle fowl (*Gallus gallus*), turned up in areas such as Southeast Asia, the Indus Valley, and China about 8,000–5,000 years ago (International Livestock Research Institute 2007, 25; Serpell 1998b). Given the numerous species of animals that populate Earth, it is noteworthy that relatively few have been successfully domesticated.[7] Moreover, "only 14 of the 40 or so domesticated mammalian and bird species provide 90 per cent of human food supply from animals" (International Livestock Research Institute 2007, 24). The big-five global food species are cattle, sheep, pigs, goats, and chickens, and it is estimated that "the world has over 1.3 billion cattle—about one for every five people on the planet. The world's sheep population is just over one billion— one for roughly every six people. There are about a billion pigs, one for every seven people, and 800 million goats, one for every eight people. And chickens outnumber humans by 3.5 to 1 worldwide; there are nearly 17 billion of them" (International Livestock Research Institute 2007, 24).

So why have so few animal species been domesticated? It is thought that "domestication presupposes a 'social medium.' As a rule the social evolution of a species must have reached a certain level before domestication becomes possible" (Zeuner 1963a, 9).[8] The animals that have been tamed or domesticated are those that tend to form packs, flocks, or herds.[9] Such species have a social hierarchy in which a lead animal is usually recognized. For example, cattle adhere to a "dominance–submission system"; once the lead animal has established itself as the boss, the rest of the herd fall into line. If no system existed to influence the behavior of group members, they "would spend all their time fighting and have no time for mating, rearing young, or foraging for food" (Carlson 2001, 30). This pattern of organization is conducive to the survival of the herd and the formation of social relations with animals of other species, including humans. However, not all socially organized species have been successfully domesticated; exceptions include hyenas, antelopes, and gazelles. This indicates that there may be a behavioral barrier to successful domestication (Bökönyi 1989, 24). The ideal animal candidate for domestication is a non-territorial species that "lives in large, wide-ranging herds of mixed sexes, organized in hierarchies, has a wide tolerance of different food plants, a short flight distance, and a relatively slow response to danger" (Clutton-Brock 1994, 28).[10] Other favorable characteristics include "lack of aggression towards humans, a strong gregarious instinct and willingness to 'follow the leader,' . . . the ability to breed in captivity, relatively short birth intervals, large litter size, rapid growth rate and a herbivore diet that can easily be supplied by humans" (International Livestock Research Institute 2007, 27). Cattle, sheep, goats, and pigs comply with many of these features. Gazelles, however, segregate into two separate herds, except when mating, and the males of the species are fiercely protective of their domains and will take flight at the first hint of danger (Clutton-Brock 1994).

The crux of animal domestication is thought to lie in the intentional changes to species brought about by those who control or manage the animals' breeding (Sándor Bökönyi, cited in Ingold 2000, 64). Thus, a domesticated animal has been defined as "one that has been bred in captivity for purposes of economic profit to a human community that maintains complete mastery over its breeding, organization of territory, and food supply" (Clutton-Brock 1989, 7). Juliet Clutton-Brock (1994, 27) maintains that this term should be applied only to those animals whose propagation has come under the auspices of human beings, because "the end product of domestication is the breed, which may be defined as a group of animals that has been bred by humans to possess uniform characteristics which are heritable and distinguish it from other animals within the same species." That animal participants learn to adjust to this interspecies cultural context implies that cultural adaptation is not the preserve of humans. Domesticated animals are also transformed into cultural artifacts that are socially integrated into and exchanged within human society through the mechanism of personal ownership (Clutton-Brock 1994, 27–29; Ducos 1989). The property status of animals reinforces the idea that humans are set apart from the natural world, so that "human beings, as social persons, can own; animals, as natural objects, are only ownable" (Ingold 1994, 6). Furthermore, the dominant idea that humans set out with the intention of domesticating animals has been challenged. For some, it was quite the opposite. It was only once the unintended benefits of domestication began to dawn on people that they increasingly incorporated animals, such as cattle, into their economies (Zeuner 1963b). For example, harnessing the physical strength of bulls to cultivate the land not only increased food productivity but also released some people from doing such onerous work: "The ox was the key to human socioeconomic evolution from primitive digging-stick agriculture to more energy-intensive systems which produced true surpluses of grains" (Schwabe 1994, 40–41).

The discussion hitherto has considered domestication primarily from the perspective of humans and what they stood to gain from this interspecies association. Albeit understandable, this affords a somewhat partial and overly human-orientated analysis. Stephen Budiansky provides an alternative understanding of the subject as he explores domestication from the animal's viewpoint and needs. For him, "Domestication was an evolutionary strategy not only for humans, but also for particular species of animals" (quoted in Swabe 1999, 37). He believes two species can evolve in concert through cooperation, as opposed to competition, and the argument rests on what he calls the "biological opportunity" that brought humans and animals together "to [allow] evolution to act on the biological motives of food and protection" (Budiansky 1992, 52, 62). As climatic and environmental conditions of the Pleistocene period were unpredictable, numerous animal species ran the risk of extinction. Those most amenable to change increased their chances of survival because they could more easily accommodate the volatile ecological uncertainties they faced. Organisms have two main approaches to survival. First, an environment or niche that is home to numerous species of plants and animals creates tremendous competition for resources. To survive in this context, organisms have to become highly specialized and draw every advantage from their niche. Such species are called "K competitors." Alternatively, the "R competitors" reside in less competitive environments, and their survival is based more on their ability to adapt and propagate successfully and regularly. Budiansky (1992, 64) suggests, "Most domesticated animals are descendants of the latter, the dwellers of the forest edge rather than the forest, the scavenger or grazer that can eat a hundred different foods, not the panda exquisitely adapted to living off nothing but huge quantities of bamboo. It is the edge dwellers, the opportunists, that would have been willing to frequent human campsites."

The opportunists were also biologically advantaged by their inherent predisposition toward sociality within their species, evident in their hierarchical social structure. This in turn provided a common basis for humans and animals to communicate non-verbally, particularly through their mutual interpretation of bodily behaviors such as signaling to indicate dominance or submission. It is the development an animal undergoes when growing from infancy to adulthood that holds the key for Budiansky. The persistent manifestation of juvenile characteristics into adulthood, a process known as neoteny, provides both the mechanism and biological resource to promote the maximum amount of natural variation in the shortest period of time. He notes:

> Selected as a way to adapt to a changing world, [neoteny] would have laid an even more solid foundation for the interactions of humans and other animals. The curiosity, the lack of a highly species-specific sense of recognition, and the retention into adulthood of juvenile care-soliciting behaviour (such as begging for food) of neonates would all have been powerful factors in inducing wolves, sheep, cattle, horses, and many other occupants of the Asian and European grasslands of the glacial era to approach human encampments and to allow humans to approach them. (Budiansky 1992, 80)

By and large, neotenic traits are understood to have arisen only because people encouraged them. Budiansky (1992, 107) claims that neotony is an evolutionary mechanism; thus, for him it is the other way around: "Domestication seems natural only because it happened, but it happened only because it was natural." This analysis challenges mainstream perspectives that prioritize and emphasize the central role of human intervention. He maintains that, because anthropology initially dominated this field of study, animal domestication was framed primarily as a "matter of cultural change." By inverting this assumption and proposing the centrality of evolutionary factors, he says, "Domesticated animals chose us as much as we chose them" (Budiansky 1992, 24).

Not only has the nature and extent of human–animal participation been the subject of much debate, so has the term "domestication." Some note that it is "slippery and imprecise" (Cassidy 2007, 3). Others suggest it is so problematic that it should be abandoned altogether and replaced by the concept of "cultural control" (Hecker 1982, 219). For example, Howard Hecker reckons human–animal relations can be plotted along an evolutionary continuum depending on the extent of cultural intervention, which can range from minimal cultural manipulation to maximum cultural control. This is illustrated in the shift away from the "cooperative driving of animals to specialised hunting techniques to animal keeping and selective culling to the selective breeding of animals" (Hecker 1982, 221). Hence, animal domestication is characterized by maximum cultural control and is located at the end of Hecker's evolutionary human–animal relational scale. Although his analysis captures the increasing level of human encroachment and interference in the lives of animals destined for food, "it fails to fully account for the radical changes in human social life and behaviour that animal domestication brought with its wake" (Swabe 1999, 32). To focus on the types of relationships that arose between people and particular species of animals without attending to the socioeconomic context in which these relations materialized is to omit vital "social factors that help to shape human behaviour" (Meadow 1989, 81–82).

Moreover, domestication signals a significant "change of focus on the part of humans from the dead to the living animal and, more particularly, from the dead animal to the principal product of the living animal—its progeny" (Meadow 1989, 81). Whereas hunters are interested in their prey only when it is about to be killed, social systems of livestock production sustain and protect the living animal until it is ready to be eaten (Clutton-Brock 1989). As animals come under the auspices of human care and management, this diminishes their independence and makes them ever more dependent on humans to meet their needs: hence, the significance of the adage "We take care of the animals—and the animals take care of us" (Rollin 1995, 6). This interspecies alliance is depicted as mutually beneficial and is thought to form the basis of a "social contract":

> Humans provided food, forage, protection against extremes of weather and predation, and, in essence, the opportunity for the animals to live lives for which they were maximally adapted—better lives than they would live if left to fend for themselves. The animals in turn provided food, toil, fibre, and power for humans. The situation was thus a win/win one, with both animals and humans better off in the relationship than they would have been outside it. (Rollin 1995, 5–6)

However, not everyone accepts the assumption of mutual advantage underpinning this contract. Some believe that only people stand to gain from this asymmetric human–animal coalition (Clutton-Brock 1994; see also Ducos 1989). Others grapple with the extent to which animals consent to this contract, if they do so at all. For example, Clare Palmer (1997) works from the assumption that consent does not have to be explicitly granted because it can be gained tacitly or hypothetically. Given this, she examines in more detail Budiansky's proposition that animals chose us as much as we chose them, which, on the surface, implies that animals freely entered into this contract. Unconvinced of the viability of this claim, she also notes once most animals undergo the process of domestication, there is no turning back. "Even if Budiansky is right, and historically in some sense animals 'chose' or gave 'tacit consent' to domestication, this is no longer a possible 'choice' for current generations of animals. The nature of the 'animal contract' is such that once in, it is impossible to get out" (Palmer 1997, 21). Perhaps at some level it is useful for people to believe that animals acquiesce to and mutually benefit from this contract because it justifies the power we increasingly exercise over them: "power not just over their conditions of existence, but over their very natures—and power which, with recent developments in biotechnology, is becoming increasingly absolute" (Palmer 1997, 22).

The latter point is important because it raises the following question: To what extent is the current application of agricultural biotechnology, genome sequencing of animals, and animal cloning just an extension of earlier forms of animal manipulation and modification?[11] Over the past 300 years, breeders and animal scientists have gained greater understanding of the genetic mechanisms underpinning the changes they are trying to bring about. Given that selection mechanisms have fundamentally changed since the onset of domestication, this casts doubt on whether "domestication," as an undifferentiated term, can adequately capture and represent all of these various changes (Leach 2007, 73). It recently has been suggested that, "under a four stage subdivision of domestication, the differing degrees of modification and human control can be easily accommodated" (Leach 2007, 92). For most of domestication, people have not fully understood the effects or consequences that their inadvertent and deliberate interventions have had on the animals they have modified. Bearing this in mind, Helen Leach (2007, 95) suggests that "justifying even more intensive genetic manipulation of domesticated (and yet to be domesticated) species on the grounds of ten millennia of prior experience is both unsupportable and evolutionary unsound."

All things considered, it would seem that the effective domestication of animals is as much an outcome of a "long-term process of trial and error" than anything else (Swabe 1999, 32). Irrespective of how it arose, domestication of the "big-five" animal species all these years ago has an interspecies legacy that is as significant now as it was then, albeit for different reasons. Scholars might continue to dispute the extent to which the emergence of a "walking larder" was an intentional, unintentional, or co-evolutionary food strategy, but few would query the socioeconomic, environmental, and ethical significance of people's ongoing affiliation to and increasing instrumental use of livestock animals. Such animals, especially cattle, have been the animate foundation on which personal, family, and institutional livelihoods have been built: "The emergence of the great Western cattle cultures and the emergence of world capitalism are inseparable, each feeding the appetites of the other"

(Rifkin 1992, 28). It is the rise of the Euro-American "animal industrial complex" that we now consider.

Cattle are a source of meat, dairy products, and leather. But cattle have a long-standing otherworldly significance, too. As spiritually revered animals, they have formed the basis of religious worship and sacrifice and "represented generativeness, virility, and fertility" (Rifkin 1992, 2). Clearly, humans have related to bovine animals in two main, albeit contradictory, ways: as sacred beings and commodities. Throughout history, the human–bovine relationship has crossed paths in a variety of geographical locations to form what Jeremy Rifkin calls "cattle complexes," which are "elaborate cultural networks that have helped shape the environmental, economic, and political dynamics of whole societies" (Rifkin 1992, 3). For example, the Kurgan people, characterized by infamous and unscrupulous raiding more than 6,000 years ago, played a significant role in the secularization of cattle. Rifkin regards the Kurgan as "the first real protocapitalists, transforming cattle into a vast store of mobile wealth that could be used to exert power over both people and territory" (Rifkin 1992, 28). These "Neolithic cowboys," nomadic herdsmen of the Eurasian steppes, were transformed by their ability to domesticate horses. The horse was pivotal in two ways. It increased their ability to drive large herds of cattle over vast geographical areas, and it enhanced their military capabilities. The Kurgan perceived the land they traversed as a resource to be exploited, as they had no connection to it. Albeit rudimentary, their calculative worldview ran counter to that of the agrarian peoples they invaded and conquered.

The Kurgan were also hierarchically organized: priests at the apex, followed by warriors and commoners at the base. But a fundamental tension materialized between priests and warriors over the use of cattle. The priests' role was to sacrifice cattle to their gods to secure future gifts of cattle and men. Thus, they "remained attached to the ancient economics of indebtedness and gift-giving. They viewed their health, wealth, and well-being as gifts bestowed by the gods" (Rifkin 1992, 32). The warriors' role was to instigate war to commandeer cattle. The rationale was twofold: to ensure a sacrificial offering to the gods and to accrue communal wealth. However, warriors gradually realized they could accumulate cattle wealth for themselves. This destabilized the hallowedness of ancient economics and contributed to the rise of a more secularized form of economics. "In the process, the bovine metamorphosed into a new creature. . . . The cow, once a god, was slowly transformed into a commodity" (Rifkin 1992, 32).

Perhaps unsurprisingly, the historical derivation of the word "cattle" is closely associated with the term "capital"; this etymological link is evident in many European languages. For example, "*pecunia*," Latin for money, has its roots in *"pecus,"* which means cattle (Rifkin 1992, 28). As a mobile asset, these animals were a source of food and a sign of people's status and wealth. For example, people would exchange cattle at times of marriage or wrongdoing (Carlson 2001; see also Evans-Pritchard 1940). Thus, livestock have played, and do play, such a pivotal part in people's lives that some environmental historians regard animals as "agents of historical change" (Anderson 2004, 4). For example, "A number of studies have accorded livestock an instrumental role in helping Europeans to establish colonies in other parts of the world. Especially in North and South America, the success of those imperial endeavors depended on the migrations of people *and* animals" (Anderson 2004, 4). Virginia Anderson's fascinating account illustrates how intercultural tensions between "native Indians" and colonists were significantly framed by the lack of control settler farmers exercised over their animate property. This was compounded because the indigenous people had no concept of, or category for, living chattel. Unsure what to make of these "strange beasts," they integrated livestock into their life world by seeing and treating them as "fair game." Such linguistic-cultural differences may seem trivial, but they are highly significant, especially since settlers avidly subscribed to and were keen to propagate the belief that "farming with animals was one important hallmark of a civilized society" (Anderson 2004, 8). As cultural carriers of this worldview, the settlers considered themselves beacons of civility and their husbandry practices the necessary corrective to the lack of such developments in America. By being shown how to look after domesticated animals, "native Indians" would acquire wealth, "advance along the path of civility, and eventually convert to a Christian faith that considered human dominion over animals to be divinely ordained" (Anderson 2004, 9).

But when settlers arrived in America, they were so busy getting established, because they were starting from scratch, they had little time to tend to their animals. Thus, their cattle and pigs had to fend for themselves as they roamed freely and extensively. The extent of free-range husbandry varied between the Chesapeake and New England colonies.[12] Chesapeake settlers prioritized their time and labor to grow and cultivate lucrative tobacco crops, so much so that even before the 1650s they had relinquished any diligent attempt to look after their animals. Released from restrictive husbandry regimes, docile livestock became increasingly unmanageable and vicious. Stray animals formed "wild gangs" as they reverted to less disciplined ways of functioning, a far cry from the civilized image settlers had been seeking to portray. Despite reneging on their duty of care, and despite the fact domesticated animals were becoming ever more feral, these animals were still their property. In contrast, New England colonists established themselves in towns as opposed to in remote rural locations. This fostered more formal and informal stewardship of, and control over, each other's animals. Despite this, they were unable to replicate the level of husbandry they had practiced in England because they still had to apply themselves to clearing and developing the land around them.

From the time these "strange beasts" set hoof in the New World, they caused havoc and confusion for the indigenous people. Marauding herds of unbridled livestock substantially ravaged Indians' gardens and crops as they trespassed across their land. As livestock mingled and foraged alongside deer, native hunters struggled with the idea that they could kill only deer; the other animal was out of bounds. Of course, livestock were injured and killed, sometimes as an act of reprisal for the devastation they had caused. This stirred up significant outrage among colonial farmers, because this was their property. Reaching a workable solution to this troublesome state of affairs was only half the story. The heated exchanges also revealed fundamental tensions in terms of how settlers and indigenous peoples envisaged the future and their place in it. By the end of the seventeenth century, the population of settlers and livestock continued to swell; this led to further encroachment and the eventual seizure of "Indian" land.[13] The role of domesticated animals in this colonizing mission had changed. No longer vehicles for and symbols of civilization, they played a more active role in physically clearing indigenous peoples off their territories. "In a real sense these creatures, even more than the colonists who brought them, won the race to claim America as their own" (Anderson 2004, 11).

America was a mixing pot of cattle cultures. However, three main cultural influences would greatly shape the cattle-ranching culture that emerged in the American West: Hispanic, British (especially Scots and Irish), and West African. Each cultural group also brought distinctive attitudes that contributed to how it interacted with its animals. For example, Hispanics tended to perceive cattle primarily as a means to an end, a commodity. Although Celts and West Africans shared this attitude, they also admired and cared for their cattle; bovines were esteemed animals (Carlson 2001, 86–92). Moreover, Carlson draws on the work of Professor Terry Jordan to suggest ranching did not stem from western America; it was "a cultural activity that immigrants carried out of the eastern Mediterranean and Nile as they swept out across Europe and Asia" (Carlson 2001, 64). And despite their enduring cultural resonance, the massive cattle drives through extensive grazing terrain persisted for little more than two decades, from the end of the Civil War to the mid-1880s.[14] Increasing speculation and substantial levels of British capital investment during this period fueled a cattle boom at the beginning of the 1880s. "American frontiersmen and cowboys cleared the way for westward expansion, [but] it was the English business class that provided much of the financial muscle to turn an outback into the richest and most profitable pastureland in the world" (Rifkin 1992, 52). Herbert Brayer (1949, 97) spells out the scale and significance of their intervention and suggests that this contribution should not be underestimated:

> From numerous small operations [the British] organized the great [cattle] companies; they made possible the stocking of the ranges to a degree never before attained; they invested in the best stock they could secure and imported the finest pure-bred Shorthorn, Hereford and Angus bulls to breed up the herds then on the ranges; they improved their range lands by developing water facilities, reseeding pastures, and fencing to prevent overgrazing; at a time when annual winter losses were high they introduced winter feeding on a mass scale and constructed livestock shelters.[15]

Despite this, the influx of investors and ranchers keen to get in on the boom put undue pressure on grazing territories. This also hastened the fencing off of open rangeland, because more and more ranchers grazing more and more livestock generated increasing conflicts between ranching and farming neighbors. At first, wooden fences demarcated land boundaries, but this proved to be too expensive for more expansive areas. The introduction of barbed wire in 1873 revolutionized fencing technology and allowed vast swathes of land to be enclosed (Cronon 1991). Livestock were now physically barred from mixing with other people's animals. This enabled ranchers to improve the quality of their breeding stock, and herds, and it prevented their animals from picking up diseases (Carlson 2001). However, by 1886 the speculative cattle bubble had burst as oversupply and diminished ranges caused a collapse in prices. The slump was aggravated by exceptionally bad blizzards, which led to the death of millions of cattle (Cockburn 1996). Ironically, the barbed wire contributed to the high death toll, because many of the animals became ensnared and entangled in the impenetrable fencing. The "'Devils Rope' had proven its impassability" (Carlson 2001, 110).

The Texas Longhorn cattle breed emblematic of that period was also responsible for the prolific and devastating transmission of Spanish cattle fever, a parasitic disease caused by ticks.[16] Although not a new disease, it created much conflict between ranchers and farmers once cattle speculators started to shift cattle north of Texas toward new markets in the East, especially Chicago (Carlson 2001).[17] For example, during the Civil War, ranchers in southern Texas were unable to access their "markets in the Caribbean Islands and the slave states of the southern Mississippi Valley," which greatly inflated the number of cattle in this area (Cronon 1991, 218). When hostilities ceased, vast numbers of Texas Longhorns roamed freely in the vicinity of San Antonio. Speculators, many of whom were Southern migrants look-ing for an opportunity to secure their future, realized they could cash in on this relatively worthless animate resource by seizing and branding the animals and sending them to Chicago, the "Great Bovine City of the World." Joseph McCoy, a livestock dealer, answered their prayers by procuring and creating a stockyard in 1867 adjacent to the rail station in Abilene, Kansas, that con-solidated this speculative venture and ushered in a "new livestock economy linking Chicago and the Great Plains" (Cronon 1991, 218). As Jeremy Rifkin (1992, 70) notes, "It was in this primitive frontier town that the north–south and east–west byways of America first intersected, connectmg the four coor-dinates of a great continent. Cattle stepped over the divide and onto the rail-cars, thus changing the course of America's history."[18]

But the new rail links threatened the interests of Midwestern "cattle kings," such as large-scale cattle grazers in Illinois, because western range-fed cattle were much cheaper to produce. Escalating land and taxation costs also made it less advantageous to waste land by feeding cattle on these grazing pas-turelands (Cronon 1991, 222). However, these very same rail links connected two abundant and significant agricultural resources: free grass and surplus corn (Rifkin 1992, 94). This signaled the emergence of a new approach to pro-ducing and finishing cattle—one that differentiated the rearing and fattening stages of the production process and introduced a new division of labor into stockers and feeders. Ranchers on the Great Plains became stockers, breeding and rearing cattle until they reached twenty-four months of age. Midwestern producers purchased the steers and fattened them for slaughter with corn. Clearly, increasing specialization augmented productivity. But it also relieved feeders of the more time-consuming and risky reproductive stage, so they could ensure "the corn their 'feeders' consumed would go into meat and fat rather than into inedible bone and other less marketable body parts" (Cronon 1991, 222). The age of the feedlot system dawned. Steers in the final stage of their productive lives were now placed in large, outdoor holding and feeding pens to reach slaughter weight as quickly as possible. Within forty years, feed-lots had become the principal approach to finishing cattle in Illinois and Iowa (Carlson 2001; Cronon 1991, 222).[19]

The injection of British capital undoubtedly helped to consolidate this nascent productive approach, as did the growing market for corned beef in England during the 1870s. The demand came from middle-class and upper-class groups, who had a particular penchant for fatty beef, and from the army, which provided soldiers with twelve ounces of meat per day (Rifkin 1992, 95). As an industrializing country, with a burgeoning urban working class to feed, England had already made the most of prime Scottish and Irish pas-turelands to cater to its insatiable appetite for beef (see Williams 1976).[20] But even this left a shortfall. Importing cattle from the continent was an obvious way to supplement the deficit. But during the 1860s, cattle from Europe was devastated by anthrax. Even though Britain activated very strict quarantine regulations to protect the nation's herds, the disease came into the country via Ireland and spread profusely (Brayer 1949). Given the cumulative implications of these socioeconomic factors, and animal disease, Britain looked farther afield to America.

The first transatlantic shipment of live cattle arrived in Glasgow in 1868 (Perren 1971). Such ships transported 600–1,000 head of cattle and were manned by about eighty crew. The voyage could take up to a fortnight, and the conditions on board, for man and beast, were at best rudimentary and at worst deplorable. Infections were rife as animals were crammed into damp and grubby surroundings. Because shippers were recompensed for animals that survived these arduous trips, even exhausted and ailing beasts were kept on all fours so they could stagger off the ship on reaching their destination (Carlson 2001, 125). However, "London and Liverpool soon became the chief reception centres, because these ports were already well equipped to handle large numbers of European and Irish cattle respectively, as well as being adjacent to south-eastern and north-western centres of population which were ready markets for the beef" (Perren 1971, 432). Seven years on, cattle were not only shipped to Britain "on the hoof," but also "on the hook" (see also Perren 1985). This would prove to be a revolutionary development. The inventor John Bates of New York identified a way to keep cattle carcasses cold in refrigerated rooms. In 1875, he went a step further and successfully shipped meat to England. This generated much attention and secured the backing of financiers. Before long, numerous shipping companies had followed suit, and fresh meat was regularly steaming across the Atlantic to British ports. As Rifkin (1992, 88) notes, "Refrigerated shipping established the all-important technological link needed to fashion a Euro-American cattle complex for the coming century."

This new knowledge was harnessed and applied to America's rail networks, too. Most significantly, Gustavus Swift, a New England farm boy who had become an experienced butcher, meat dealer, cattle buyer, and meatpacker in Chicago, was inspired by the level of profit to be made if cattle could be transported on the hook.[21] The main stumbling block to this venture was the problem of decay and finding a way to evenly chill meat for the duration of lengthy journeys. Swift and Andrew Chase, an engineer from Boston, created the Swift-Chase car in 1878, which was the state of the art in refrigerated rail technology. They overcome uneven cooling by locating "boxes filled with ice and brine at both ends of the car, venting them so that a current of chilled air constantly flowed past the meat" (Cronon 1991, 234; Horowitz 2006; Rifkin

1992).[22] This had the effect of nationalizing beef, as people no longer had to buy locally produced or recently slaughtered animals. Consumers took time to embrace the idea that it was safe to eat meat from cattle killed up to a week earlier. However, chilled dressed meat was cheaper than fresh meat products, which seemed to make it more palatable. Clearly, this threatened the interests of local butchers and slaughterhouses throughout the country; many protested against these developments, and many went out of business. But Swift was an astute operator because he "realized that it would be better to have the wholesale butchers as allies than as enemies, so in many towns he approached the leading butcher—usually a person of considerable means—about becoming a partner in the dressed beef business" (Cronon 1991, 241). Meatpackers' growing stranglehold on the market strengthened Swift's position to negotiate favorable deals with those who were amenable to working with him. For the first time, Americans were consuming more beef than pork; by 1909, America "had become a beef-eating nation" (Horowitz 2006, 32).

This dietary transition was hugely indebted to the pork-packing hub in Cincinnati, dubbed "Porkopolis" in the 1830s because it "pioneered the manufacturing techniques that would transform Chicago meat-packing" (Cronon 1991, 228). It was in America's hog-slaughter plants that the disassembly line was born. In conjunction with an increasingly specialized division of labor, this provided the technological template for the mass-production assembly lines that would transform steel and car manufacturing in the early twentieth century. It also ushered in the industrial mantra that "time is money," as packers became more and more focused on finding the most efficient and cost-effective ways to dismantle animals into their constituent and marketable parts. The sheer number of animals processed in Cincinnati at that time meant that "even body parts that had formerly been wasted now became commercial products: lard, glue, brushes candles, soaps" (Cronon 1991, 229). However, the blueprint for industrialized slaughter was truly epitomized by the Chicago Stockyards. The sheer scale and level of efficiency achieved at these yards marked the inception of slaughter factories. According to Daniel Pick (see also Sinclair 1906):

> With the development of the Union Stockyards of Chicago in the 1860s, . . . mechanised animal butchery moved towards its apotheosis. . . . In their specialised division of labour, the new slaughterhouses exemplified many of the points elaborated in Taylor's *Principles of Scientific Management*. The abattoir was to become a complex factory involving a precise separation of tasks and mechanical operations— stabling, killing, cleaning, refrigerating, transporting, inspecting, preparing foodstuffs, and so forth. (1993, 180)

Improved transportation links, the emergence of feedlots, and refrigeration all culminated in the Chicago stockyards. It was "the ultimate meeting place of country and city, West and East, producer and consumer—of animals and their killers" (Cronon 1991, 212).

The increasing interdependence of grain farmers, stockers, feeders, butchers, and meatpackers brought into being a powerful new agricultural business alliance that has gone on to fundamentally alter meat-eating habits in America and beyond. The animate basis of these significant socioeconomic, cultural, and dietary changes can be traced back to the intended and unintended consequences of domesticating animals. Clearly, this interspecies legacy is an ongoing process that has formed a "domesticatory system" characterized by two main elements: the production and consumption of animals (Digard 1994, 233). Having outlined the rise of a Euro-American cattle-meat economy and the democratization of meat consumption in Western societies, I now consider how the taken-for-granted nature of the social contract between traditional farmers and their animals was undermined, especially in the United Kingdom after 1945, as livestock-production methods became increasingly intensive. In so doing, this takes us to the practical and ethical cornerstone of productive human–livestock relations: the changing role and status of stockmen and what is believed to constitute good stockmanship.

During and following World War II, the United Kingdom, other European countries, and the United States pursued a productivist agricultural policy that led to the industrialization and intensification of livestock production and increased meat consumption in these countries (see, e.g., Kim and Curry 1993; Winders and Nibert 2004).[23] A practical implication of this policy was a reduction in the number of stockmen employed to look after an increasing population of livestock (Rollin 1995). This changed the nature, opportunity, and frequency of interactions between stockmen and their animals. For example, talking to the animals and "praise with the voice," which encouraged the naming of individual animals, was undermined by increased stocking levels (English et al. 1992, 34). Production animals were less likely to be regarded as individuals. More fundamentally, the knowledge and technology now existed to alter both the animals and their productive environments to enhance their overall productivity. This meant that animals could be kept in contexts that were not conducive to their biological and social functioning. As Bernard Rollin (1995, 137) notes, "Industry replaced husbandry—with the help of new technology, one could meet the select needs of animals that were relevant to efficiency and productivity without respecting the animals' entire telos or psychological and biological natures." For instance, putting unprecedented numbers of pigs and poultry into intensive housing systems would have been practically impossible without the assistance of veterinary personnel and the animal-feed and pharmaceutical industries (Pye-Smith and North 1984; Swabe 1999). The routine administration of antibiotics and vaccines to manage the increased incidence and risk of infectious disease was pivotal to rearing animals in these conditions. But, John Webster (2005, 100) suggests, "It is an unequivocal insult to the principle of good husbandry to keep animals in conditions of such intensity, inappropriate feeding or squalor that their health can only be assured by the routine administration of chemotherapeutics."[24]

This major transition from husbandry to industry did not go unnoticed or unchecked in the United Kingdom. The publication of Ruth Harrison's *Animal Machines* (1964) was so successful in raising the British public's awareness of and concern for intensively produced livestock that the government was forced to set up a committee to inquire into the welfare of animals kept in such circumstances. The publication in 1965 of the *Report of the Technical Committee to Enquire into the Welfare of Animals Kept under Intensive Livestock Husbandry Systems,* by F. W. Rogers Brambell, was very significant because "it was the first formal recognition by an official body that intensive animal agriculture raised animal welfare problems" (Garner 1998, 152). In effect, the Brambell report signaled an end to the relatively unquestioned nature of the age-old relationship that had existed between farmers and their animals. Human–livestock relations had become problematized, politicized, and subjected to increasing public and scientific scrutiny.

Prior to this transition, subsistence livestock production was thought to be epitomized by good husbanding skills, the practical basis of the "social contract." The extent to which traditional farmers adequately met the physical and biological needs of the animals in their care would have a direct bearing on how well the animals developed and their level of productivity. By respecting these interests, both stood to gain from this so-called partnership: "The animals' interests were the producers' interests" (Rollin 1995, 6). However, the preference for more extensive husbandry approaches was not a result of benevolence; farmers were simply adopting the most pragmatic strategies that best suited their agricultural circumstances and requirements (Webster 2005, 98). For instance, pigs and poultry were allowed to be semi-independent, wandering around the farmstead and rummaging for suitable things to eat, because this supplemented their upkeep and made them fairly low-maintenance animals. Moreover, it was the amount and quality of land that ultimately determined the number and types of livestock animals to be produced on traditionally run farms. So it was common practice to produce an assortment of livestock, as opposed to concentrating on one species (Swabe 1999, 119). But as livestock production shifted away from sustainable small-scale farming toward more intensive units, the link between land and level of production was severed. "Most or all of the inputs to the system—power, machinery (e.g., tractors) and other resources (e.g., food and fertilisers)—are brought in so that output is constrained only by the amount that the producer can afford to invest in capital and other resources and the capacity of the system to process them" (Webster 2005, 98). Such changes also had an impact on stockmanship, a key but underappreciated productive factor.

Stockmanship has been defined as "a human activity that applies the ability, knowledge, skills and common sense necessary in optimising health, welfare, husbandry, management and thereby both physical and financial performance, in animal production" (Benyon 1991, 67). But stockmanship has also been described as "the world's most undervalued profession" (English et al. 1992, x). These quotations indicate the ambiguous status of stockmen. They are absolutely central to the productive process, yet their contributions have not always been formally acknowledged or appreciated. Until recently, the livestock sector focused most of its attention on designing and refining the constituent elements of animal-productive contexts: "genetics, nutrition, housing, health, management and administration." Stockmanship was the "forgotten pillar" in this productive process, and the contribution of stockpeople to that process has tended to be overlooked (English et al. 1992, n.p.). A key reason for this oversight is that "stockmanship has had no champion to promote its importance" or financial sponsorship to fund research and development (English 2002, 3). However, the industry is realizing that any system of production is only as good as the stockpeople who work within it (Hemsworth and Coleman 1998).

Since the early 1970s, stockmanship has been ascending agricultural and animal-science research agendas. For example, Martin Seabrook found that dairymen who brusquely handled their cows got lower milk yields than did more confident handlers with introverted temperaments (see, e.g., English et al. 1992; Hemsworth 2004). Similarly, it was found that poorly handled pigs not only produce higher levels of cortisol, a stress hormone, which "adversely affect[s] growth and reproductive performance by disrupting protein metabolism and key reproductive endocrine events" (Hemsworth 2000, 4), but also can become very wary of people. Animal handlers who shout or move erratically can also alarm livestock and induce fear. Paul Hemsworth (2007, 198) suggests that "opportunities for positive human contact are probably reduced in modern production units . . . [given that] many routine husbandry tasks undertaken by stockpeople may contain aversive elements." This is an important point. On the one hand, those working with livestock have to understand how their temperaments, attitudes, and actions can affect the animals' lives they attend to. This requires a degree of reflective and caring practice. On the other hand, stockmen are expected to carry out potentially aversive husbandry practices, such as castration and dehorning, because it is part of the commercial process of cattle production (see, e.g., Rollin 1995; Stookey and Watts 2004; Webster 2005). Thus, cultural-structural expectations and legal requirements also shape the actions of those charged with looking after livestock. Ideally, workers perform husbandry practices competently to ensure that animals experience minimal, if any, discomfort. Given that some interventions can induce varying degrees of pain, agricultural workers may have to carry out tasks they would rather not have to do. These types of situations may generate structurally induced ambivalence for those working in such contexts (Merton 1976; Wilkie 2005). However, since many handling studies seem to pursue too narrow an understanding of the stockmanship role, this can mean that non-productive challenges associated with the role, such as its emotional, cognitive, and ethical aspects, have been virtually factored out or ignored.

In recent years, the livestock industry has been keen to improve how stockmen handle livestock because it enhances animals' well-being and productivity. This realization has become a key commercial driver in terms of encouraging research into the human–livestock interface. Such studies have been carried out in intensive productive contexts, such as dairy, pig, and poultry units, as opposed to more extensive systems, because there are increased opportunities for contact between producer and livestock (English and McPherson 1998a).[25] Much attention has also been paid to the breeding end of the productive process rather than the finishing stages, for similar reasons. This means less work has been done on human–beef cattle interactions, although there has been some exploration of farmer–veal calf relations (e.g., Lensink et al. 2000). Researchers are also keen to develop ways to understand how stockmanship and related practices are experienced from, and affect, the animal's perspective throughout its lifespan (e.g., Boivin et al. 2003; Waiblinger et al. 2006). Clearly, these types of studies provide a useful understanding of the practical and economic impact of human–livestock interactions. However, some note:

> The existence and influence of empathy between stockperson and farm animal is generally, although not universally, appreciated in agriculture. Some authorities in trying to define "good stockmanship" consider that it consists solely of using technically correct methods in handling and managing farm livestock. The impact of the emotional relationship between stockperson and animal is therefore largely discounted in such definitions. (English et al. 1992, 35)

Good stockmanship is especially valued for its instrumental role in producing economically viable livestock and enhancing animal welfare. Thus, much of this research has focused on identifying additional ways to maximize the productive effects of this interspecies relationship. Cultivating empathy might be the next element to be harnessed by the industry to contribute to this aim. But adopting such a calculative perspective, irrespective of how understandable it may be, diminishes the stockperson-animal interface to little more than another productive factor that needs to be understood so it can be fully exploited. Not only animals are increasingly controlled and manipulated in productive contexts; so are those who handle them. The commercial livestock sector is highly competitive, so producers are acutely aware of the economic benefits they stand to gain from promoting an animal welfare-friendly image; it enhances animal productivity, facilitates product sales (which increases profits), and improves the overall public image of the industry.

Some think it is just a matter of time before the industry and the general public "place an increasing emphasis on ensuring the competency of stockpeople to manage the welfare of livestock" (Hemsworth 2004, 34). Thus, identifying what constitutes good stockmanship is becoming ever more important as the sector reviews how best to recruit, train, and retain suitable employees (English et al. 1992). But there is a growing sense of urgency about this very issue, because the standard of British stockmanship is believed to be under threat from "low profitability of farming leading to reductions in staffing to reduce costs, and the breakdown of succession on family farms as new generations seek different careers. Over time these factors will diminish the knowledge, skills, and experiences of British stockmen that are essential for high standards of welfare" (Farm Animal Welfare Council 2007, 1). In 2001, it was estimated that 292,000 stockmen worked in the United Kingdom, but given that this type of work is perceived to be of low status with unsociable hours and poor working conditions, the sector has its work cut out if it is going to attract high-quality candidates (English and McPherson 1998b; Farm Animal Welfare Council 2007). As the traditional pool of farm labor dries up, future workers may have minimal, if any, firsthand experience working with livestock. How important will this change be, considering that some within the industry query the extent to which good stockmanship skills can be formally acquired?

Traditionally, sons and daughters on family farms have acquired their husbandry skills by serving a long and informal apprenticeship following in their parents' footsteps. This began as soon as they were old enough to walk: "The children were almost automatically trained in all aspects of stockmanship from the time they progressed from crawling to toddling. They thus developed their rapport with the animals from a very early age. Their awareness of animal behaviour, of animal needs, and their handling skills developed through their own experience and via the instruction and guidance of their parents" (English et al. 1992, 5–6).

A senior auctioneer who has worked closely with farmers and stockmen over many years believes that good stockmanship skills are handed down from one generation to the next:

> A stockperson more often than not they've been born to it. You'll get the exception where somebody comes in, out *o* [of] the city or whatever, and does the job, but you'll find that most of those folk are born to it; they've been bred to it. You'll find that their father was [a] good stockman and grandfather was a good stockman or something, way back. If there *wisna* [was not] that trait in their makeup, they *widna* [would not] be in it.[26]

Another auctioneer thought a good stockman was someone "who sees an animal that's slightly off colour the day before it becomes ill" (see also Garner 1946), while others, who should know, "won't notice it till it's probably too late." For him, "It's something that he has, and he just knows. And others would work a lifetime in it and never have it." However, he seemed unsure about whether such skills were "bred in you" or "learned from a very young age." Clearly, it is important to differentiate early exposure to and experience with livestock from the notion of genetic inheritance, because informal experiences socialize and instruct a person in the ways of handling livestock. Nonetheless, a common thread running through these accounts is that good stockmen are perceived to inherit their practical skills and knowledge from their forefathers, and this cannot be short-circuited. The notion of a genetic metaphor was also evident in John Gray's ethnographic study of hill sheep farmers in the Scottish Borders. He found sheep farmers understood the social transmission of personal attributes, farming attitudes and skills in terms of a biological mechanism:

> These attributes are said to be "bred into" a person, that is, acquired and transmitted through genetic processes such that they inhere in and are consubstantial with the person's body. The metaphorical quality of this understanding is that farmers' knowledge of the genetic mechanism of sheep breeding gained in their everyday farming activities is used to conceive of the process through which people acquire and transmit attributes. . . . In such a theory the aptitude for farming is corporeal such that personhood and farming

> co-mingle in a common substance that, *like* sheep's embodiment of
> the hills and farming people, appears to be transmitted genetically.[27]
> (Gray 1998, 354)

Two recurring statements capture this biological cosmology: "Farming is bred
into you" and "good farmers come from farming families." These sentiments
seem to resonate with my respondents' beliefs about good stockmanship, but
the degree to which husbandry skills might be inherited or passed on through
early socialization is clearly open to debate. Perhaps farmers involved in
producing and reproducing high-quality livestock are more likely than finishers to draw on biological analogies to make sense of their everyday lives.
The breeding histories of livestock, especially but not exclusively pedigree
animals, are crucial to assessing the potential quality of the progeny they produce, so why not metaphorically apply this worldview to assessing the caliber
of those who look after livestock, too?

It is worthy of note that stockmanship is becoming officially valued just
as the taken-for-granted agricultural reserve from which competent livestock
workers is derived is shrinking. This adds a further dimension to the afore-
mentioned nature-or-nurture debate, especially in light of the demographic
shift that has occurred over the past 100 years or so. As mentioned earlier,
where more than "half the population made a living producing food, this
has dramatically changed. Today barely 1.7 percent of the population works
in production agriculture, with perhaps half of that group or less in animal
agriculture" (Rollin 2004, 7). This begs the question of who will be interested
in working with livestock if they have no previous experience in doing so. An
interesting observation that emerged during my research was that more than
three-quarters of all my female interviewees had experience working with
horses; half came from a farming context, and half did not.[28] Perhaps the
knowledge, skills, and confidence gained from handling large animals such
as horses give women with little, if any, experience with productive animals
a way into a male-dominated industry. Further, most non-farming entrants
are probably more used to interacting with companion animals. If this is the
case, it will be interesting to see how they adjust to working in a commercial context where animals are more instrumentally regarded and routinely
slaughtered.

So what elements are believed to constitute good stockmanship, and what
is meant when someone is described as a good stockman? In recent years,
there have been copious examples of bad husbandry practice in the industry.
For example, animal-welfare and animal-rights groups have done much to
highlight the darker sides of food-animal production and slaughter-related
practices in their various exposé documentaries, literature, and campaigns.

Although such coverage is justified, less attention has been paid to examples of proficient animal handling. This imbalance contributes to the belief that many workers in the industry intentionally or unintentionally mistreat animals. Of course, some animal-advocate perspectives start from the absolutist assumption that rearing and killing livestock for food is morally wrong and totally unnecessary. Even if producers treat their animals well, this does nothing to assuage their primary objection. This is a consistent and justifiable position. Alternative (and more mainstream) standpoints within and outside the animal-welfare community reckon that it is morally acceptable to eat animals as long as their welfare has been given due consideration. These are diverse and competing viewpoints, but they have a common basis: Most of the commentaries are made by non-farming people located outside the industry. Such perspectives have significantly shaped our understanding of and discourse about what is good and bad in terms of contemporary livestock production. However, we are far less informed about how those who actually work within the various animal-production sectors make sense of such matters. A useful starting point is the "three essentials of stockmanship" identified by the Farm Animal Welfare Council (2007, 7):

> *Knowledge of animal husbandry.* Sound knowledge of the biology and husbandry of farm animals, including how their needs may be best provided for in all circumstances.
>
> *Skills in animal husbandry.* Demonstrable skills in observation, handling, care and treatment of animals, and problem detection and resolution.
>
> *Personal qualities.* Affinity and empathy with animals, dedication and patience.

The ability to relate to livestock or have "stock sense" is crucial. For instance, a cattleman claimed that he could "almost think like they do." When moving cattle, he often got a sense of "when they might think there's an opportunity that they could run away from you." At times, he was so tuned in to a beast that he was "a split second ahead of that animal" just because he was "thinking along the same lines."

An estate manager said that, when hiring stockmen, he prioritized the person's knowledge, affinity, and ability to handle animals. One way to establish whether a candidate had these valued characteristics was through "word of mouth." He explained: "The shepherd we head-hunted because we knew by reputation that his parents were excellent stock folk and we head-hunted him from another job to come and work with us. The cattleman . . . came from the college farm; again he came by reputation, he was known as good, he was a good stockman." A female farmer also said she relied on the influence of "hearsay." In addition, she asked how long a candidate had worked for his previous employer: "Relevant practical experience, many years of working with stock on the same place for a long time would be a good indication of the worth of a man, as far as livestock's concerned." The farm estate manager also elaborated on what he meant by the term "affinity": "It means that they can instantly recognize an animal that's not thriving: The animal is not well. They can relate to animals when they're calving, if they're having difficulties, how long they leave them, some you leave longer than others. . . . They're quite happy to stay up all night to make sure the cows have calved or sheep [have] lambed. They often put the livestock before themselves; they can work long hours."

Good stockmen have been described to me as "gold-dust," "first-class," and "absolute specialists" (see also Garner 1946). Others have said that good stockmanship is a "way of life" and "a calling," as opposed to just a job. A recent attitudinal survey of 516 dairy farmers in the United Kingdom found that they valued the following personal and farming skills in a good stockperson: "temperament (69%), knowledge of dairy cattle (57%), interest in dairy cattle (51%), conscientiousness (including commitment to working long and antisocial hours) (49%), love of cows (44%) and it is something one is born with which cannot be taught (11%)" (Bertenshaw and Rowlinson 2009, 64). A dedicated stockman may be highly valued, but the social and personal cost of such diligence should not be overlooked. A farm manager explained how he sometimes felt quite low because he was so "tied down," especially if things were not going well: "Although you like working *wi* [with] animals it can get you down that you've *tae* [to] get up at four every morning and trail out every night." This level of commitment disrupts his family life, too. For example:

The wife will go away *wi* my son *tae* football or something, and I *canna* [cannot] go because I'm watching a cow that's calving or something, or going to calve. She's showing signs that she'll probably calf that afternoon or something, so you've got to stay about here. . . . I might sleep on the couch and just pop out and in every hour, set *ma* [my] alarm for every hour. . . . [I]t can disrupt your social life. I mean, I *widna* really say I have much of a social life, probably because *o* that. I *canna* really have, I don't really have hobbies.

The Farm Animal Welfare Council is particularly concerned about those who work on their own because of the "effect of stress on the welfare of stockmen, whether caused by isolation, excessive hours of work, lack of succession, ill-health or financial concerns, and consequently on the welfare of stock under their care" (Farm Animal Welfare Council 2007, 10). Since the livestock sector in the United Kingdom is characterized by an aging cohort of farmers of whom 60 percent are sole operators, this indicates how the welfare of both humans and animals can be closely interlinked.

However, being described as a good stockman is an important accolade; it is a source of informal prestige and status for those working within the industry. For instance, the cattleman observed that, since agriculture is not the best-paid industry, "If you *dinna* [don't] like what you're doing, I really don't think you'd stay. So you have to like what you're doing. . . . It certainly *isna* [is not] the money that keeps you here, but interest, doing well in your job." For him, "interest" is about taking pride in his work and trying to better himself. The cattle grazing in the fields are there for all—especially fellow farmers and stockmen—to see. He explained, "A lot of pride goes into your work, 'cause you know that other people are looking at it." This might occur informally when farmers "take stock" of and gaze at neighbors' cattle while driving, at times too slowly, through the countryside. An old Aberdeenshire maxim is that "you do your best farming by the road." Members of the general public may also appreciate seeing livestock dotted throughout the rural landscape, but they are less likely to cast a productive eye over their physical conformation. So it would seem that, for producers, the cattle's physique comes to symbolize and reflect the standard of their husbandry skills while the animals are alive, and when dead, the quality of their carcass. Hence, the animals are the raw material through which stockmen can visually display their practical skills, build up or undermine their professional reputations, and create their sense of self. These types of observations are perhaps more likely to occur in and apply to small- to medium-scale cattle-production systems that still have an extensive element. For example, "natural" beef from suckler cow herds are calves reared by their mothers that are then finished on pasture. However, about a half of all beef production in the United Kingdom and northern Europe is derived from calves born to the dairy industry. As these calves are intensively produced, they tend to spend their entire lives confined indoors on concrete slats and are fed artificial diets (Webster 2005, 154). This type of production, known as "barley beef," first arose in Britain during the late 1950s, following the outcome of animal-nutrition studies carried out by the Rowett Research Institute (Brambell 1965, 42). The use of slatted floors has always been controversial because they are physically uncomfortable for the animals and contribute to painful feet and limbs. A government review accepted that slats are useful for removing dung but recommended, ideally, that "where animals are permanently housed indoors an alternative form of flooring should be available to them" (Brambell 1965, 46, see also Webster 2005).

Clearly, good stockmanship can be beneficial to the industry, to the animals, and to those who practice it, albeit for different reasons. However, what it means, and takes, to be a good stockman depends on a range of factors, such as the species of livestock being produced; the scale and type of the productive context; the knowledge, skills, and personal qualities of the stockman; and who is making that judgment. The ethical basis of the social contract between traditional farmers and their animals has undergone substantial change over the past sixty years or so, to the point where the coincidence of animals' interests and producers' interests is more implausible in contemporary large-scale animal-production systems. There has been a growing trend toward producers' interests increasingly eclipsing animals' interests, although some producers will continue to respect and prioritize the needs of their animals. Despite the burgeoning research into the nature of human–livestock interactions, too much emphasis seems to have been placed on the instrumental benefits of cultivating this relationship purely for economic reasons. Even though the commercial rationale for pursuing this line of inquiry is understandable, it fails to take adequate account of how stockmen make sense of, and deal with, the different relationships they build up with the animals with which they work each day.

It is tempting to privatize and separate emotional concerns from more public-oriented economic concerns (Anderson and Smith 2001). But as the following chapters illustrate, my data indicate that instrumental and emotional components of livestock production can and do coexist. Since many of the handling studies have been carried out in breeding units, and good stockmanship is commonly associated with the reproductive stage of the productive process, this does not fully address the types of challenges faced by workers who are responsible for finishing animals for slaughter. Because new recruits to the industry are less likely to have been socialized into working with productive animals, such issues are surely of increasing importance and policy relevance. Perhaps future research into human–livestock relations can benefit from exploring the full range of people's physical, cognitive, emotional, and ethical relations with breeding, store, and prime animals, "because this more accurately reflects the ambiguity of the various productive contexts within which those at the 'byre-face' negotiate their everyday practice" (Wilkie 2005, 228).

3

Women and Livestock

The Gendered Nature of Food-Animal Production

> *Old Macdonald had a farm, ee-eye, ee-eye oh*
> *And on the that farm he had a cow, ee-eye, ee-eye oh*
> *With a moo, moo here and a moo, moo there*
> *Here a moo, there a moo*
> *Everywhere a moo, moo*
> *A quack, quack here and a quack, quack there*
> *Here a quack, there a quack*
> *Everywhere a quack, quack*
> *Old Macdonald had a farm, ee-eye, ee-eye oh.*[1]

"Old Macdonald's Farm" is a popular children's song that portrays a highly nostalgic and romanticized image of traditional livestock farming. It conjures up a small-scale family farm where the farmer dutifully attends to his farm animals. The farmer is assumed to be male, and farming is presented as men's work. Even today, the commercial production, marketing, and slaughter of livestock, especially cattle, continue to be predominantly, although not exclusively, a man's world. However, the term "family farm" juxtaposes the personal and the public and brings into focus the interdependent nature of these two normally distinct, and largely gendered, spheres of modern social life. I show that women have played, and do play, a significant, albeit relatively unacknowledged, role in livestock-related farming contexts. By exploring in more detail why women are less likely than men to work with cattle, and the justifications put forward by my respondents to explain this pattern, I suggest that not only the production process but also the animals are gendered. Overall, this chapter provides an insight into a relatively under-explored aspect of contemporary agricultural life and an opportunity to think about how stereotypical notions of masculinity and femininity are expressed, justified, and contested in a range of commercial and hobby livestock-productive contexts. I start this discussion by providing a historical overview of women's roles in farming.

Women have been, and continue to be, integral to the everyday running of small to moderate-size farms, but until the 1980s they were largely absent from agricultural research studies.[2] When women were acknowledged, it was primarily in their role as farmer's wife, mother, helper, and assistant (e.g., Brandth 2002a; Rosenfeld 1985). Up to the mid-eighteenth century, work on subsistence family farms could be divided into three areas: the fields, the barnyard, and the household (Osterud 1993, 19). The fields were primarily a male sphere of responsibility, but women would help during busy times, such as the harvest. The barnyard was a domain commonly worked by men and women, indicating that livestock farming seems to rely more on the joint input of men and women.[3] Domestic and child-care chores associated with the household were predominantly women's work. This fairly typical sexual division of labor reflected the physical arrangement of the farm. In the main, "Women's labor centered on the house, men's work on the fields . . . [and the] two met in the barnyard" (Neth 1995, 19). The reality was often messier than this neat schema. Although late-nineteenth-century agricultural reports in America depict working in the fields as "men's work," women often joined men. For example, "It was appropriate for women to work in the fields . . . during periods of labor shortage, in particular crops, or if they were black, immigrants or poor" (Sachs 1983, 20). The production of cotton and tobacco in the states of the U.S. South readily used, and was highly dependent on, black women's labor. However, if at all possible, white women's rightful place was considered the home. Given this racialized dimension, "The desire to keep white women out of the fields is not based on the presumption that women cannot perform agricultural labor; it is a matter of status for white men that they can keep their women in the home" (Sachs 1983, 25). Farmers' wives and daughters were thus a versatile and crucial reserve of farm labor that could be drafted in during busy times of the farming year. This practice has been exemplified by a historical study carried out by Kathryn Hunter and Pamela Riney-Kehrberg (2002), who explored the type of farm work young women did in Australia, New Zealand, and the Midwestern United States during the period 1870–1930. They note that "rural girls were workers, although the exact nature of that work varied with family composition, location, stage of settlement, crop mix, and a number of other factors, particularly the availability and affordability of hired labour" (Hunter and Riney-Kehrberg 2002, 136). From an early age, boys and girls were socialized by their parents into the importance of working hard and acquiring the knowledge and skills relevant to their farming contexts. Young women could, and did, on occasion turn their hands to most tasks, but gendered assumptions influenced and demarcated the type of work thought to be most appropriate for them.[4]

Violating such assumptions could earn social disapproval. For example, in the summer of 1898, the New Zealand *Otago Witness* published a letter from fourteen-year-old "Jessie C" that outlined her contribution to the family farm (Hunter and Riney-Kehrberg 2002, 140–141). She milked cows, drove sheep, managed a team of horses, and was on the verge of learning how to plough. She added, "I have only killed one sheep yet, but I intend to kill more" (Hunter and Riney-Kehrberg 2002, 140). The last comment in the letter sparked a thirteen-month debate that would have gone on for longer if Dot, the editor of the section, had not decided to stop printing letters on that issue. So what was the problem? Quite simply, "Killing a sheep, not to mention feeling a sense of accomplishment for having done so, was beyond Dot's comprehension of acceptable female behaviour" (Hunter and Riney-Kehrberg 2002, 140).[5] Other readers also expressed their displeasure at the sheep-killing incident: A young women called it "unwomanly, and not at all a thing to boast about," and a boy objected to Jessie's apparent pride in her ability to kill a sheep by asserting, "I think it is very out of place and unmaidenly" (Hunter and Riney-Kehrberg 2002, 140).

Other readers thought Jessie's behavior was acceptable and framed their support in an equal-rights discourse while others asserted a notion of usefulness. Proponents of the latter position were more pragmatic in their response. They were aware that it was far from ideal for farm girls to kill sheep, but they recognized that there might be occasions when it was quite simply unavoidable. One woman wrote, "It shows a girl's pluck when she can kill a sheep. . . . [I]t is not work for a girl, but sometimes a girl has to do such a thing. I don't think it a ladylike accomplishment, certainly; still I admire her pluck and nerve." Another girl wrote, "Things on a farm would soon become mixed if the girls began to say that certain things were not girl's work. I have killed a poor calf to put it out of pain, when there was no one else to do it, and I don't consider myself hard-hearted" (Hunter and Riney-Kehrberg 2002, 141).

Thus, even though men and women were both involved with the management and husbandry needs of livestock on family farms, a division of labor emerged that revealed a distinct pattern: "Work with certain animals and specific tasks, such as milking, herding, collecting eggs, feeding, and butchering, [we]re often gender specific" (Sachs 1996, 103). Furthermore, during the middle of the nineteenth century, the "cult of domesticity," an influential middle-class, urban Victorian ideology, permeated farmyards and homesteads throughout Western Europe, Australia, New Zealand, and the United States (Hunter and Riney-Kehrberg 2002; Shortall 2000). This ideology advocated a gendered spatial division of labor. As societies industrialized, women became responsible for unpaid productive and reproductive labor in the domestic sphere, while men colonized the public domain of paid productive work. It is interesting to consider the etymological meaning of the term "domestic," given that it was initially defined in the Oxford English Dictionary as "having the character or position of the inmate of a house; housed" (Sachs 1983, 47). Carolyn Sachs (1983, 47) contends that the use of words such as "inmate" and "housed" is indicative that "domestic labor is forced upon the worker by others [and that] women's domestic duties do not simply involve work that must be done in the home, but seem to imply confinement to the home." An important implication of this gendered ideology for farming households was that it created uncertainty because "rural societies were increasingly unsure how their daughters should negotiate the slippery slope between necessary labour and desirable domestic feminine accomplishment" (Hunter and Riney-Kehrberg 2002, 135). This is particularly relevant to family farms because in this working context, the boundary between work and family is far from clear (Rosenfeld 1985). So where previously farmwomen had been valued for their ability to roll up their sleeves and carry out a wide range of non-domestic farming tasks, they increasingly faced the challenge of straddling and negotiating two coexisting and, at times, contradictory versions of acceptable femininity.

This was compounded by the transition from subsistence to commodity production, which contributed to the de-feminization of two previously female-dominated spheres of food-animal production on the family farm: dairy and poultry farming. It was typical for farmwomen to be responsible for the daily upkeep of chickens and dairy animals and the byproducts derived from these animals: eggs, milk, butter, and cheese. Such activities gave them, albeit to varying degrees, the ability to secure a degree of financial independence, or "pin money," that in turn enabled them to contribute significantly to the subsistence needs of the farm household in times of economic hardship (Sachs 1996). However, as men increasingly entered and took over these productive spheres, the relatively independent socioeconomic niches carved out by farmwomen receded, as did their links to productive work within farming (Brandth 2002a, 188). Given the historical and socioeconomic importance of these two feminized areas of food-animal farming, I now consider them in more detail.

The farmer's wife or dairymaids primarily oversaw dairying activities on small and large farms in England (Pinchbeck 1969).[6] These women were entrusted with milking the cows, making butter and cheese, and selling the products at local fairs and markets. Farmers would frequently consult dairy-women about the amount and standard of the milk being produced, as well as about the animals themselves and their general upkeep (Pinchbeck 1969, 11).

Although dairywomen had a great deal of influence over the livestock and the processes and byproducts associated with milk production, they rarely owned these resources. In fact, because land ownership and inheritance of farms was in the main a patrilineal affair, farmwomen were routinely bypassed and thus systematically excluded from acquiring key economic resources. If, however, there was no male heir, the farmer's widow or daughter would acquire and have ultimate responsibility for the farm (Pinchbeck 1969, 10). Even though dairy work was arduous and unrelenting, it was a rare example of esteemed farmwomen's work: "It was one area of work where women did receive recognition, status, income and a certain degree of power" (Shortall 2000, 248).

In the second half of the nineteenth century throughout Western Europe and North America, dairying underwent major changes as it industrialized to become more commercially viable. A common characteristic of these transformations was that dairy work became progressively de-feminized (Shortall 2000). Even though farmers' wives and daughters were increasingly displaced and replaced by male dairy workers, the pattern did not occur uniformly across different countries.[7] For example, a comparative study of dairying in New York and Sweden between 1860 and 1920 shows that, while farmers' daughters in New York responded to the changes by renouncing their involvement in cheese making, young farmwomen in Sweden actively participated in the industrialization of the dairy industry and embraced the economic and personal opportunities it gave them (Sommestad and McMurry 1998). These women continued to perform dairy-related work because of "the strong feminine coding of milk" that permeated and typified the ethos of small-scale dairy farms throughout Sweden, especially in the north, and because the labor carried out by these women had been essential to the continued existence of these dairy farms (Sommestad and McMurry 1998, 150–151). Long-standing gendered taboos had effectively dissuaded men on Swedish farms from milking: "Not only was it improper for a man to milk, it was considered shameful" (Sommestad and McMurry 1998, 150–151). In a related vein, attempts to secure acceptable milk yields on preindustrialized Norwegian dairy farms, located in climatically challenging environmental contexts, depended on dairywomen's perceived ability to draw on magical powers and supernatural folklore. The inhospitable natural conditions set limits on the level of milk production, which contributed to the potency of, and belief in, supranormal powers to mediate the uncertainty of dairying at that time (Skjelbred 1994, 199–200).

Such folklore, however, was largely absent from, and less likely to shape, dairying practices in New York, and it was more acceptable for men in that region to be affiliated with milking. In addition, by the 1860s any residual associations linking notions of "female nurturance and the milking of cows"

increasingly had been swept aside as cows were progressively reconceptual-
ized as machines (Sommestad and McMurry 1998, 152). This mechanistic
construction of the human–dairy cow relationship directed attention toward
procedural and technological issues that identified ways to refine milking
techniques, which in turn reinforced the need for more systematic attempts
to efficiently manage and sanitize the dairy-production process. For example,
in 1903 the Rockefeller Foundation and the New York City Department of
Health conducted a study of dairy sanitation that led to Sarah Belcher's *Clean
Milk*. The publication explained how milk-production methods, as envisaged
and informed by the emergence of agricultural scientists and public-health
professionals, aimed to produce milk devoid of bacterial impurities. This sanitized, urbanized, and de-feminized vision of industrialized dairying was symbolically encapsulated in a new image of "a man in a clean white uniform, his
head covered with a clean white cap, milking cows in a pristine environment"
(DuPuis 2002, 125). The once ubiquitous and romanticized image of dairymaids milking cows now represented a lack of scientific progress. In effect,
the productive roles, economic contributions, and relatively prestigious status
of dairywomen were progressively curtailed as scientific and technological
developments recast and redefined dairy work as legitimate contexts for male
labor.

Similar changes affected poultry farming. Before 1940, women looked after
chickens. One of the daily chores of the farmer's wife or her children was feeding the chickens and collecting eggs. This made sense because poultry roamed
freely, were in close proximity to the homestead, and could survive on leftover
food scraps (Sachs 1996). Given the feminization of human–poultry interactions, such animals were located at the bottom of the food-animal hierarchy
and were taken less seriously by male farmers. In contrast, "The male sphere
of farming [in America], which included work with cattle and horses, was tied
to the symbolism of the cowboy and the bravery and strength of conquering a
frontier" (Fink 1986, 136). Despite such a disparaging and dismissive attitude,
a study of rural women in Iowa clearly illustrates how chickens really proved
their worth during the Depression. The egg money accrued by farmers' wives
prevented many farms from collapsing at a time when the agricultural products produced by men, such as cattle and corn, provided little or no family
income (Fink 1986, 142). The U.S. agricultural census indicates that in 1930,
98 percent of farms in Iowa kept chickens, with an average flock of 169 birds
(Fink 1986, 136). This pattern of small flocks overseen by female farmers was
pretty much replicated throughout the United States (Sachs 1996). However,
by the 1940s, moves were afoot to expand, rationalize, mechanize, centralize,
and intensify poultry farming. Consequently, chicken produce became separated into two distinct and highly specialized systems of industrialized food-

animal production: meat and eggs (Dixon 2002; Ellis 2007a; Kim and Curry 1993; Sachs 1996).

This historical synopsis offers some insight into the transient nature and gendered legacy of farm-animal production systems. Although it is beyond the remit of this chapter to characterize all of the possible cultural, regional, and gendered permutations that human–livestock systems can take, there does seem to be a recurring pattern that is worthy of note.[8] As farms increase in size and become ever more industrialized, the way work is allocated also changes. Carolyn Sachs (1996, 104) suggests that factors such as the species and economic value of the animals and the scale of production tend to influence who is in charge. As a rule of thumb, men oversee large species of livestock produced on a large scale for meat. Women care for young and small animals reared for non-meat products such as eggs, milk, and wool. The spatial location of animals and their proximity to the household is also a key factor. For example, in a recent study carried out in Scotland, it was suggested that tasks associated with looking after animals can quite often be broken into "small units of work that can be fitted in with other commitments such as running the home" (Schwarz 2004, 114). Farmwomen are expected to fit these animal-related chores into their existing domestic responsibilities. However, it is suggested that the age and needs of young children, combined with the positioning of animals in relation to the farming household, may greatly influence the extent to which farmwomen can indeed carry out such tasks. Finally, not only are men's and women's involvement in these systems different; their involvement also seems to be hierarchically organized and valued—that is, men producing cattle generally attract more socioeconomic prestige, rewards, and status than women producing sheep or poultry.[9]

In my study, just over 40 percent of the women were hobby farmers ($N = 9$), of whom only a third were born into a farming background.[10] In contrast, all of the women employed in the livestock auction ($N = 5$) came from a farming background, as did more than three-quarters of the farmers ($N = 4$). Of all of the women who had participated in further or higher education, veterinary personnel ($N = 4$) consistently achieved the highest level of educational attainment, tended to be younger than the rest of the women, and remained single. Finally, more than half of the hobby farmers owned their farm or smallholding, as did three-quarters of the farming group. However, most of those who owned their farm or land were married. My hobby contacts were small-scale, part-time producers who tended to keep rare breeds of livestock; all except one couple worked with sheep, and only a few women had experience in owning cattle.[11] One of these women owned two animals, and one had twelve (this included a bull, and she was awaiting the arrival of calves). Interestingly, one of the male hobby farmers experimented unsuccessfully with cattle when

	CATTLE	DAIRY	SHEEP	PIGS	GOATS	RABBIT	POULTRY
TABLE 3.1 NUMBER OF ANIMALS IN DIFFERENT TYPES OF PRODUCTION SYSTEMS							
Hobby	0–12	N/A	14–60	11–25	3–22	15	70–1,000*
Organic†	4	145	51	15	N/A	N/A	140
Commercial	90–1,000	N/A	300–1300	N/A	N/A	N/A	N/A

Note: N/A, not applicable.
* Includes rare and commercial poultry breeds.
† Two male organic farmers were in my study; one was a commercial dairy farmer converting to organic milk production, and the other was an organic vegetable farmer who also produced livestock.

he bought six calves over a period of time. He found them costly in terms of veterinary bills and, more important, felt that he "didn't have the facilities . . . for dealing with these damn great beasts" and decided to sell the cattle at a loss. Other types of livestock commonly kept by the hobby farmers included pigs, goats, angora rabbits, and poultry. Thus, sheep and certain types of small livestock are far more prevalent than cattle. This supports survey findings by Nick Evans and Richard Yarwood (1998) that revealed a species preference among keepers of rare breeds: 59 percent owned sheep, while 39 percent owned cattle. Unfortunately, their survey did not provide a gender breakdown of the Rare Breed Survival Trust (keepers) membership. Controlling for gender would have indicated whether female and male keepers preferred sheep or cattle, respectively. Table 3.1 gives an indication of the stocking levels for different species of animals owned by my participants involved in hobby, organic, and commercial production.[12] The pattern of stocking levels tends to support the observations made by Sachs (1996).

Shepherdess Rather Than Cowgirl: Why Do Women Tend to Work with Sheep?

From the accounts that follow, it becomes evident that some women are wary of cattle because of their size and weight. Interestingly, wariness also applies to rams. For example, two female hobbyists who preferred working with sheep declared that they were in awe of cattle, which made them feel less confident when around them. A hobby sheep breeder maintained that she was "petrified of cattle," even though cattle had never kicked her and she had not had any other bad experience. However, she did enjoy working with a friend who owned a herd of Dexter cattle, a small Irish breed, but declared, "I wouldn't ever really be at the stage that I could say that I would have a bull." She also acknowledged that the size of bulls was an issue but then realized that she was afraid of rams, too. "I do prefer sheep," she explained. "But having said that, I don't have a ram, for the fact that I don't think I could handle it." She recounted that the first year she had hired a ram for mating, she had tended to put a halter on him; even though the ram was "a good character," she said, she had been told that he could be "quite a bully." Overall, her dealings with the ram had been benign, but when she walked through the field, she found that he would actually go for her, especially at mating time. She concluded, "Although I'm . . . confident with sheep, I'm certainly not with rams."

Another female hobby farmer who kept Norfolk Horn sheep proclaimed, "Sheep are wonderful," and that she could easily "empathize" with them. She had been drawn to Norfolk Horns when, while leafing through a book that depicted rare livestock breeds, she came across a picture of a ram that "had magnificent horns." She also explained that the big spiral horns were surprisingly useful when handling the animals, and "they add more masculinity to a male sheep, these great big things, it's just their masculinity—they looked good." She also believed that sheep have "no malice," even though her husband had been butted by a ram. (She made sense of this negative human–animal episode by assigning the blame to her husband.) Her Highland cow also had "huge beautiful horns," she said, but she was not convinced that she "could have a relationship with her." She described her relationship with the cow as one of "intimidation": "She's just not friendly. She just didn't play the game. If she's got her head in a bucket of food, you can touch her. Forget it otherwise." A Highland cow, she noted, "is basically a wild animal," whereas a Jersey cow they had once owned, and that her husband had particularly liked, was "tame" and "would come to you." She explained the Jersey's tameness as the result of daily milking.

A former sheep farmer who had produced commercial breeds also felt intimidated by cattle. She recounted a situation in which she had helped a farming neighbor castrate his male animals. "I felt threatened," she said. "I didn't feel comfortable, and I didn't feel that I had the confidence that was required to work with them. I'm sure they wouldn't have ever done anything, but if ever I had to go out to them, for whatever reason, just to walk through them . . . there was always in the back of my mind that one was going to run me down." She identified two reasons for thinking this way: their size and because she "hadn't actually had anything to do with them." Even though she considered herself relatively safe because her friend knew his animals well, she continued to watch her back. She feared she could not cope "if for some reason they got out of control [and the situation] became dangerous. I wouldn't have known what to do, where to go. . . . In my mind I might not be standing up to do anything about it anyway." Although she remained vigilant and respectful of rams and newly lambed ewes, she was much more at ease with sheep. She alluded to a couple of instances in which rams had been "nasty" to her but did not impute the malign motives to them that she did to cattle. "I was in [the rams'] way, and they wanted to get out, and they just could see the exit and they just wanted to go for it. But that wasn't aggression." Rams could "display their power and strength if they want[ed] to," she said, "but the ewes, not at all. Again, I've only been biffed by ewes when they [were] protecting their young, and I would have done that to somebody to protect my young. I've never perceived them as a threat."

Finally, a small-scale breeder selected the Ryeland breed because they were "medium-size sheep which ladies could manage to handle." She liked the breed because they needed little care and attention and were straightforward to lamb, and, she noted, their fleece was valuable. She added, "From the 'pretty' point of view, . . . they're nice to look at." Because she found them aesthetically pleasing, "like a cuddly toy in a shop," she preferred Ryelands to Rouge and Suffolk breeds, which she considered less physically attractive. Another Ryeland sheep breeder also noted that they looked "like a teddy bear," but she also appreciated their placid disposition. Recounting a rare-livestock show, she said, "It was a Jacob's show and all the Jacobs were going absolutely bananas, like they do, and this little Ryeland lamb stood chewing his cud and I thought, yes, that's what I want, something that looks like a teddy bear and behaves absolutely phlegmatically in all states of crisis."

In summary, for the reason of size, female hobby farmers I interviewed saw cattle as potentially dangerous and hard to handle. Although they acknowledged that rams could also be threatening, especially during breeding seasons, they were more manageable because they were much smaller. They were also viewed more charitably: Their hostile actions were excused by focusing on the human involved in, or the context of, the incident. Finally, even ewes' occasional outbursts were framed to sound less threatening. Words such as "biffed" play down the risk posed by ewes; further, safeguarding their lambs provides an acceptable and fully justified rationale to explain ewes' aggression.

These accounts also indicate a gendering of the animals that goes beyond making distinctions of gender. There is clearly a physical dimension to the way the female farmers thought about threat. In both species, the male is larger and more aggressive than the female. However, because all cattle are potentially more dangerous than all sheep, the ranking of the two genders of the two species in a danger list that runs bull, cow, ram, ewe in effect means that the continuum acquires a gender gloss. The bigger, more aggressive animals are "male"; the smaller, more docile ones are "female."[13] It is interesting to note, however, that male workers are also very respectful of bulls. For example, an experienced livestock handler who worked in the mart noted that he "never trust[ed] a bull" because he believed they could "turn on ye [you] in seconds," and because of the "weight o them, yer goosed [i.e., you have a serious problem]."

Furthermore, different breeds of livestock are allocated their place on the male–female continuum according to size and reputation. Experienced male mart workers noted that cattle breeds such as Aberdeen Angus, Friesian, and Red and Black Herefords had earned a reputation of being "quiet beasts" that are easy to handle. In contrast, Limousins, a continental cattle breed, are considered "a *bitty* [bit] flighty *kin'* [kind]," which leads workers to be more "wary *o* them." Of course, individual animals may deviate from the imputed characteristics of their breed. Animals expected to be docile may turn out to be quite fiery, and vice versa. Thus, the location of animals on this continuum is not fixed. Personal experiences vary, as might the level of perceived threat, but these factors feed into how men and women come to regard and relate to the animals with which they work.

The feminization of sheep—their smaller size, docility, and apparent lack of malice—also contributes to their being considered, by men and women alike, safer and easier to handle and thus more suitable for women to look after. For example, female livestock handlers working with cattle in the livestock auction market tended to deal with the paperwork, while women working with sheep were hands on, physically moving the sheep from their pens to the sale ring and back. An auctioneer who specialized in marketing sheep commented that women were more likely to be involved with sheep at lambing time. He thought their input was helpful because they showed patience, an attribute he thought women were better able to offer. He contrasted this with the physical nature of working with cattle but acknowledged that some women were involved in the cattle side, albeit less so. He believed that greater prestige and status were attached to working with cattle than sheep because cattle were economically more valuable than sheep and pigs. Interestingly, a farmer once asked him, "When are you going to move on from sheep to become a real auctioneer?" Auctioneering remains a male-dominated profession; this question therefore suggests that being closely associated with sheep, a highly feminized animal, undermines not only a man's professional status as an auctioneer but also, possibly, his masculinity. The inverse is also possible. In other words, women can potentially increase their personal status and standing in the (male) farming community by working with cattle, more masculine-coded animals.

For example, a hobby farmer explained that she had regarded herself as a "real farmer" when she had cattle, but now that she had only sheep, she no longer did so. She thought her "street cred" had been higher among commercial farmers when she had worked with cattle and drew this conclusion from the "terrific amount of interaction" she had had with other farmers while working in an agricultural retail business and attending the local livestock market. She noted that farmers would say, "Oh yes, you've got a few sheep, have you?" If she said, "Oh, I've got a cow," she got quite a different reaction. She attributed the difference to many farmers' assumption that "any *wifie* [woman] can look after sheep. There's a very, very different perception of somebody with cattle." She was also aware that big farmers in the area who produced sheep commercially were regarded seriously by fellow farmers, in contrast to people who had migrated from England or other areas of Scotland and kept a few sheep. They, she said, could be seen as "a bit of a joke." Such "incomers" also tended to be quite affluent and subscribed to romanticized perspectives of country life. This indicates that relationships between "incomers" and "locals" can be strained and a source of resentment (Jedrej and Nuttall 1996). Since northeastern Scotland has a reputation for producing beef cattle, my respondent thought that Aberdeenshire farmers perceived cattle as both a "traditional farm animal" and a "serious animal." Cattle were "fairly big; they're fairly strong; they command fairly big prices; they need handling; [and] they wrap up an awful lot more money. The price of one beast compared with the price of a sheep is in a different league entirely. . . . Sheep don't command the same respect up here as cattle do." If you owned cattle in this area, she proposed, you entered the "real farmer's world," which was a "man's world." Although there were "some very good women cattle breeders up here, they're not the norm. . . . There are very few women bidding for cattle. I mean, the women will be at the mart, but they'll be having their coffee. . . . Except on rare breed days, there are very few [women] there."

Handling livestock, especially cattle, can pose a serious occupational hazard; such animals can be unpredictable and very dangerous. Thus, personal and institutional concerns about possible and actual risks to the health and safety of workers are well-founded (Kouimintzis et al. 2007; Lindsay et al. 2004): "Internationally, farming is considered to be one of the most dangerous occupations, with the risk of injury being approximately 5–10 per 100 workers per year" (Dimich-Ward et al. 2004, 52). Moreover, the report *Exploring the Influences of Farming Women and Families on Worker Health and Safety* (Health and Safety Executive 2006, 1) noted that "agriculture has the highest fatal accident rate of all industry sectors [in the United Kingdom]. In 2002/03 the rate was 9.5 fatal injuries per 100,000 workers. . . . [I]t also has the highest prevalence of occupational ill health in comparison with other industries." Among a range of possible factors that contribute to the high incidence of farming-related accidents, a "change in cattle breed" was identified as a key issue (Health and Safety Executive 2006, 8). Canadian research identified that men and women experience different patterns and types of agricultural injury (Dimich-Ward et al. 2004). For instance, men are more likely to be injured by machines, and such injuries can be life-threatening. Since tractors and farm machinery have been coded as masculine, men are more likely to use such machinery and are thus more subject to machinery-related injuries (e.g., Brandth 1995; Leckie 1996b; Saugeres 2002b, 2002c). By contrast, women's injuries are rarely life-threatening and are often caused by animals.

Livestock-auction workers are especially vulnerable to injury because they handle a transient population of unfamiliar animals. Auction managers alert to such risks prefer to employ experienced stockmen if they can, precisely because they are skilled at handling livestock, especially cattle. But recruits entering the industry increasingly do not come from farming backgrounds and may have little, if any, experience working with cattle and sheep. This trend has prompted the mart to appoint a training manager to prepare novice animal workers for this challenging interspecies environment. Despite formalized attempts to address the emerging drain in stockmanship skills, some managers are inclined to think that there may be limits to what can be expected of inexperienced workers. For example, one manager explained that it was crucial to ascertain the exact nature of people's experience with livestock before assigning them to work with cattle or sheep. "There's no substitute for experience," he said, particularly if they are working on the "cattle side of the sales." He was very keen to attract "farm workers or farmers that are looking for a bit of hard cash," but, he said, he "would never put a woman in a pen full of cattle. . . . Calves, OK, that's fine, but you've got to have a happy medium, and they do realize that we've got some really good strong lad[s] down there, and there's just no one can take their place." In contrast, he noted, anyone could work on the "sheep side of things," even if he or she had never handled livestock before, because "it's a total[ly] different job. . . . [W]e just need people to chase sheep, you know, drove them to the sale ring and drove them away from the sale ring again, which anyone can do as long as they can walk. That's all they need to do, really."

A hierarchical division of human–animal labor seems to operate in the auction system: Cattle require experienced male farm workers, especially male stockmen and farmer's sons, but anyone, regardless of sex or experience, can work with sheep. Because the mart is the public face of the commercial livestock industry, a high premium is placed on workers who have the knowledge, skills, and confidence to handle animals well. This is particularly relevant to those who are responsible for parading animals around the sale ring, because that is front-stage work and open to public scrutiny and evaluation. The handlers are as much on show as the animals. Since male workers are charged with this job, the public face of the industry continues to be represented by men.[14] As a manager explained, those who do this high-profile job have to be "very presentable . . . because it's our face [in] the ring. . . . If there's any filming to be done it'll be done in the ring." Further, he suggested that there was no occasion to hire women for the ring, as "the guys that do it have done it for years and years and years." Another manager was disinclined to put women in the sale ring out of concern for their physical safety: "If the beast [cattle] turned . . . a man *wid* [would] maybe *hae mair* [have more]chance *o* getting *oot* [out] *o* the road." Similarly, a mature female livestock handler who had a farming background and experience working with cattle regarded "*gaan roon* [going round] *wi* the cattle in the ring" as men's work. "I don't think we'd be quick enough *tae* get *oot o* the *wye* [way], we *widna* see danger the same." Even during calf sales, where there might be minimal, if any, concerns about workers' safety, male handlers ushered the young animals round the sale ring. Sometimes they provided their fingers for the calf to suck, which it did, and then the animal followed the handler around the ring. Meanwhile, another worker sometimes stroked the calves gently in the pen as they were being prepared to enter the ring.

Even though women are not present in the sale ring and would not be expected to work "in a pen full of cattle," they are deemed capable of managing calves backstage in the byre. For example, a female worker was assigned this job when male workers who previously had handled the calves were perceived sometimes to lack patience; the manager wanted to find someone who would be more respectful to the calves. When he discovered that the woman liked feeding calves and enjoyed working with them, he invited her to take over this animal duty. Fully aware that "it was *aye* [always] the farmers' wives that fed the babies," she simply replicated a long-standing gendered tradition that defined the looking after of newborn and young livestock as "women's work."

This gendered dimension to, and division of, human–livestock labor appears to be based on an assumption that men and women have or lack different skills, attitudes, and beliefs in terms of handling different species and ages of livestock. For example, "Progressive feedlots, slaughter plants and auctions are hiring more women to handle animals. The managers report that they are often gentler and more careful with animals" (Grandin 2000, 3). Similarly, a French study found that female veal farmers were more likely than male colleagues to express the importance of positive and empathetic attitudes and behaviors toward the animals they produced; this was evident in their more considerate interaction with the calves (Lensink et al. 2000). Because practicing empathetic stockmanship skills is thought to enhance animals' welfare and increase their productivity, such outcomes would be valued by the commercial livestock sector (English et al. 1992).

Attending to the needs of newly born, young, and sick farm animals has traditionally been the role of the farmer's wife. My respondents continued in this tradition. For example, a commercial cattle farmer who has 250 breeding cows and finishes most of her calves for slaughter, as well as sheep, said that men and women deal with livestock in very different ways:

If my husband is injecting an animal, he'll just use any old syringe. If I'm injecting an animal, I'll make sure that the syringe is clean before I use it. He wouldn't ever bother to go that little extra distance to save getting an abscess from a dirty needle. I would always try and make an animal [as] comfortable as possible if it was ill, and by that I mean getting a bit of extra bedding for it, getting a bucket of water for it if it can't rise, making sure that it has food in front of it. If that animal is ill and off its feet but is able to eat but it's in a court with other animals . . . I'm the person that sits—literally sits on the ground—and makes sure that only the sick animal eats the food that's in front of it. My husband would never do that. [He] doesn't have the patience for that. That's my job. I know that I have to do that; otherwise, that animal could be neglected. Or it could take longer to recover because it's not getting food. Also at lambing time, I'll always look after the pet lambs because I will clean the implements, the bottles and the jugs that are used for making up the milk. I'll keep the milk at the same strength, at the same temperature, [and] I'll feed them at the same time. Whereas [my husband] would [say], "Oh, that'll do! I'll feed them now." No—feed them when they're meant to be fed. You know, they're babies, and they like routine. So oh, yes, we've different approaches entirely.

Later in the interview she felt the need to qualify her "strong relationship" to her animals by suggesting that if she had children, the potency and significance of these interactions would probably diminish. In the absence of motherhood, she said, "I feel this need to look after things. . . . So probably that's why I have quite a strong relationship with certain things that have been ill or unwell. . . . But don't get me wrong. At the end of the day, I'm still thinking, I've turned it around, I'm going to get some money from you at the end of the day. There's always that goal, always."

This financial reality check is possibly her attempt to offset the intense bonds she has formed with her animals. In a line of animal work in which it would be fairly unusual for male farmers to overtly admit such feelings, she is clearly deviating from the tacit emotive-cultural norms that govern this sector. By playing up her commercially orientated credentials and emphasizing the economic imperative that drives and underpins her actions, she realigns herself with these key norms. This recurring theme of making money is highly gendered because making money is equated with "proper" or "real" farming, whereby "real" farmers are assumed to be male. This farmer seems to be aware that, if she wants legitimately to claim the masculinized title and status of "farmer," she needs to de-feminize her account and play down her strong relationship with her animals.

Similarly, a female sheep farmer who once had sixty or more breeding ewes also felt there was a difference between how men and women managed livestock, especially in terms of cleanliness and handling pregnant and birthing animals. She said that men were less attentive to sanitation and justified the statement by claiming that women are naturally endowed with "a better ability to judge things with regards to hygiene.... [W]e're going to be the main carer for offspring, and nature's made us such. Out there in the lambing shed, I know if it had been a man doing it, I think the losses would have been much higher, because I really don't think they would have been as conscientious with regards to sterilizing." Her preoccupation with cleanliness and limiting the incidence of cross-infection was very much "drummed" into her during her nursing career, and the professionalized knowledge she acquired through that training diluted her essentialist assertion that such knowledge is innate to women. Regardless, she seemed to insinuate that women have a superior understanding of and relationship to, in this case, expectant animals.[15] She also believed that having "relatively small hands" lessened the physical trauma she might cause ewes while lambing. Although she conceded that male farmers do generally "manage their animals very well," when it comes to assisting birthing animals, she said, "they're much, much, rougher." She justified her capacity to be more empathetic with pregnant animals by noting that she had also given birth: "I've got a lot more empathy because I am female and I've given birth, and I know what it feels like. I know all the emotions that are involved and how your hormones are, you know, and maybe as a female there is more empathy involved than with the male, because no man's given birth yet, to my knowledge."

This farmer drew on her embodied knowledge of and experience with childbearing and childbirth to inform and legitimate her allegedly more respectful, gentler, and more empathetic approach to lambing ewes. By implication, this narrative excluded all male farmers and all women who had not given birth. The tendency for women to "mother" livestock also was evident in some of the auction workers' accounts. A woman in her thirties who had five children, for example, regarded the animals she worked with (mainly sheep) as her "bairns" and thus felt nurturing toward them. She recounted a situation in which an orphaned lamb had been left at the end of a sale because it had injured its leg. Instead of letting it go to slaughter, she took it home and arranged for veterinary treatment. In that case, she was keen to do all she could to alleviate any discomfort the sick animal might be experiencing, even if, at the end of the day, it would still be sold.

The "Hardy Farmer" Stereotype

Although the livestock auction remains a male-dominated arena, managers appreciate the presence of female employees because "they seem to calm situations down a lot better than what the blokes can." As one manager explained, "The old days of . . . big, sturdy and strong guys working with livestock has disappeared. . . . Instead of using your brawn, you use your brain." This shift in emphasis is driven partly by increasing British and European Union farm-animal-welfare legislation and the bureaucratic fallout generated by and associated with the BSE crisis in 1996. The official and legal terrain within which the commercial sector now operates has changed, as has the range of skills and attributes required by personnel to work in this context. For instance, older-generation farmers are finding it increasingly difficult to deal with the administration related to livestock production. In contrast, younger-generations of farmers and their wives and partners tend to be computer literate; having such skills means they could have a greater say in the running of the farming business. The balance of power may be changing, providing an opportunity for dominant patriarchal influences within the industry to be challenged and diluted. If we are witnessing a fundamental transition in livestock farming, perhaps long-standing traditional notions and expectations of what it meant to be a man working in this context will also be increasingly eroded, devalued, and redefined.

A female mart worker with a farming background said that most of the women who worked there were "animal lovers," and those who were not did not tend to last long. "If *yer* [your] an animal kind *o* person," she said, "you've got an understanding for the animals and you'll put up *wi* things *fae* [from] animals that *ither* [other] people *widna* put up *wi* because *ye* can understand why their *deein fit* [doing what] their *deein*." In contrast, her male colleagues, especially farmers' sons, would not openly admit to being "animal lovers" because they were "macho." The expectation that young men "brought up in the country [will be] tough [and] hard" has long persisted in the farming community, she noted, but "it's maybe *nae* [not] as bad as it used to be." Nonetheless, "They *widna* like *onybody* [anybody] *tae* think that they're *safties* [weak]." An auctioneer echoed this position:

I suppose a lot *o* them *dinna* want to be seen to be softhearted. They're producing meat, if the farmer or the farmer's wife has spent days trying to keep a lamb or calf living, you do get some attachment to them. You *dinna* just want to go and say well, bang, you're dead. . . .

I suppose the typical farmer is supposed to be a big burly chap, and nothing seems to worry him as such. Quite honestly, I feel a lot *o* them are very softhearted, but they *dinna* like to show it; they *dinna* like to show their feelings.

A former commercial farmer also thought it was problematic for male farmers to talk openly about any emotional attachments they might have toward their animals: "It may come down to part of the male make-up not wanting to be seen to be having feelings and emotions." In contrast, a young stockman maintained that he "would probably be much softer, emotionally wise, compared to a lot of farmers. A lot of them . . . wouldn't have emotions that I have attached to animals. They are just an animal to them." He then qualified this statement: "I'm sure the ones with the cows producing the calves seeing them from day one; they possibly would have much more emotions because the animal is there so much longer. There's a big farmer lives next door to us [who] buys all his cattle at the mart and would have them no more than a year. He wouldn't show any emotions at all. I could be wrong, but I don't think so."

The stockman worked in a breeder-feeder unit, which means that, unlike the "big farmer" who specialized in buying a transient flow of unknown store animals to fatten up for slaughter, he managed a known cohort of animals. Different stages of the production process and the extent to which stockmen follow the same animals through the process provide differing opportunities and constraints in terms of how emotionally involved with or detached from animals such workers may become. Moreover, younger and future generations of farm workers are perhaps more likely to deviate from traditional role expectations that previously shaped how older generations of farmers portrayed and expressed their manliness. The stereotype of the emotionally aloof "hardy farmer" appears to coexist with other ways to be masculine—in this case, one that allows a more overtly expressive man to come to the fore.

It has been suggested that there are two ideal types of masculinity in agriculturally related contexts: monologic and dialogic. Monologic masculinity represents "a conventional masculinity with rigid expectations and strictly negotiated performances that provide a clear distinction between men's and women's work. . . . [It] also limits the range of topics deemed appropriate to discuss" (Peter et al. 2000, 216). Livestock farmers are less likely to typify this form of masculinity than arable farmers who work with "big machines": One of Gregory Peter and his colleagues' respondents said, "You don't dominate [animals] the way you dominate land. Animals are much more humbling because they're . . . harder to control' (Peter et al. 2000, 228). Dialogic masculinity characterizes a more flexible interpretation of manliness; it "is more open to talking about mistakes, to expressing emotions, to change and criticism" (Peter et al. 2000, 216). Of course, it is unlikely that any male farmer would ever fully approximate the characteristics of either ideal type; male farmers are likely to have varying elements of both. Other research has identified three masculinized farmer-success templates, which pretty much maps onto, and has been derived from, changing historical, sociocultural approaches and norms to farming—that is, agrarian, industrial, and sustainable (Barlett 2006). This not only provides a more nuanced, multiple, and shifting understanding of farmers' attitudes, beliefs, and practices, but also militates against seeing male farmers as a homogenous occupational grouping.

The stereotype of the hardy farmer resonates outside the commercial livestock sector, too. A female hobby farmer, for example, showed how easy it is to generalize from one's personal experience, and her use of the pronouns "they" and "we" had the effect of depicting all male farmers as heartless and uncaring. She said that she was more able to acknowledge feelings for the livestock she kept than "hardy farmers," who appeared to be "outwardly feelingless." She had not met many farmers who talked "truthfully about what they feel," which made her wonder whether they experienced similar feelings but, unlike her, did not express them. She also noted that, when talking about ill animals or animals that had to be killed, men and women expressed themselves differently. Men tended to adopt a more shocking and macho tone of language. "We just took a *spad* [spade] and *chapped* [knocked] it on the *heid* [head]," she recounted a male farmer as saying. "They seem to delight in doing that, and I haven't met many women who would turn round to you and say, ah well, I *jist* [just] *chapped* that puppy on the *heid* cause it only had three legs or something like that. No, I think that's the macho thing, isn't it, with men."

Farming has traditionally been a patriarchically organized occupation in which senior male figures have undoubtedly shaped the sociocultural and emotional norms for those who work within it (Tolson 1977, 51–54). Commercial livestock production is no exception to this tradition. Monologic masculinity may well be a more legitimate characterization of male agricultural workers functioning within such a system, but working with animals potentially dilutes the purity of this ideal type, which may foster a more latent dialogical type of masculinity to be informally expressed. However, given the precarious state of livestock farming, it is particularly important to note that the "struggle to survive in farming is in part a struggle to retain one's identity as a man. Farmers who are less in control of their farm, less productive, and less successful may be considered less masculine than other farmers. One

defensive response to agriculture's uncertain structures of masculine performance is to assert a rigid, oppositional, socially controlling masculinity—a
strongly monologic masculinity" (Peter et al. 2000, 228).

It is tempting to explain emotional aloofness as just a local expression of
a general male characteristic. However, this underestimates the sociocultural
and historical context from which many older-generation agricultural workers descended: the *fermtouns*.[16] David Cameron (1995, 14) characterizes the
northeastern *fermtoun* folk this way: "It was not a polite society. Far from it.
Greeting was coarse and often abrasive. Its folk did not praise highly; few were
masters of the facile phrase or skilled in that fulsomeness that oils the wheels
of self-interest. If that made them more awkward, maybe it also made them
more honest, their friendship a finer thing." The hardy farmer is partly born
out of the austere experiences of those who were the backbone of the "poor
man's country" (Carter 1979, 1). The northeastern folk to this day are generally renowned for being *dour* (grim), reserved, and phlegmatic. Thus, overt
declarations of emotion may not be made in the commercialized livestock
sector, but this does not necessarily mean that male agricultural workers do
not develop emotional attachments, albeit to varying degrees, to some of the
animals they produce.

There continues to be a gendered dimension to food-animal production. Moreover, the level of status attached to the various tasks and activities
associated with processing and producing livestock appears to be closely
tied to, and mediated by, factors such as the scale of production; economic
value; and species, sex, age, and health status of the animals with which men
and women are involved. However, this gendering of the production process also extends to the gendering of the animals themselves, whereby cattle
and sheep are ranked by women and men along a male–female continuum
according to factors such as their size, breed, and reputation. Bigger and
more aggressive animals are "male"; the smaller, more placid animals are
"female." Locating animals along this continuum is closely related to and
informed by the extent to which those working with them perceive the animals to be a threat. In practice, the plotting of animals on the male–female
continuum is not hard and fast; rather, it is dependent on and shaped by the
nature of human–livestock interactions. Thus, the same animal may be perceived and gendered somewhat differently by different workers. Similarly,
an animal may initially be regarded as "male" but over time come to be seen
as less of a threat and thus increasingly regarded as "female." The inverse is
also the case.

Nonetheless, in my study women tended to feel most confident working with sheep, especially ewes and, to a lesser extent, rams. They were less
trusting of, and more intimidated by, male and female cattle. This pattern

was particularly evident in hobby farming, though evidence of it was found in the commercial livestock sector, too. Finally, small numbers of women from farming backgrounds have entered the livestock auction, a male-dominated sphere of agricultural work. They tend to work with sheep rather than cattle, are primarily responsible for the administrative tasks of processing livestock through the marketing system, and are employed behind the scenes. In all, this ensures that the public face of the livestock auction and the commercial livestock industry itself remains male. Given the growing shortage of skilled livestock handlers, the extent to which the industry will continue to be a male world remains to be seen.

4

"Price Discovery"

Marketing and Valuing Livestock

> Farmers find solace amid the cacophony of bleating, lowing
> and gate clanging that is the thriving auction mart. Rubbing
> besmocked shoulders with fellow farmers, comparing notes with
> dealers, sharing jokes with the canteen staff who slip whisky into
> their coffee, they feel strength in numbers there. In a profession
> changing with bewildering speed, they take comfort in the mart's
> timeless rituals. The auctioneer's banter is as much a sound of the
> countryside as is the first cuckoo. His ability to read the minds
> and interpret the twitches and nods of craggy-faced buyers is as
> much a part of rural folklore as is the shepherd's delight of the red
> sky at dusk. (Farndale, quoted in Graham 1999, 111)

The livestock auction is the public face of the commercial livestock
sector and plays a pivotal role in the marketing and pricing of
animals bred and sold for human consumption. In this chapter,
I trace the ascendance of the modern auction system, and auctioneering, in Britain during the nineteenth century and discuss some of the
components that go into valuing livestock. I also show how key players
in the early years of the fat-cattle trade, such as drovers and dealers, were
increasingly displaced and eclipsed by the growing importance of butchers, as demonstrated by the notion of the "butcher's beast" (Trow-Smith
1959). Finally, I explain why, from the mid-1940s on, the number of markets steadily declined and allowed alternative marketing approaches such
as the deadweight system to come to the fore.

Markets and fairs have been described as "the veins of the livestock
industry: without them the blood of pastoral production would cease to
run"; as such, they were generally located in key regions where livestock
was produced and meat was consumed, and any suitable place between
these two areas (Trow-Smith 1959, 226). Following the Union of the
Crowns in 1707, a thriving droving trade emerged between Scotland and

England to the extent that the mass movement of farm animals on the hoof became a highly significant component of Scotland's economy (Haldane 2002, 2). Aberdeenshire farmers had long driven and sold their beasts south. Thus, drovers were used to accompanying their animals over fairly long distances to major cattle trysts that, like that of Falkirk, were routinely convened "on the same day of the year at each place" (Thomson 2005, 9). Unsurprisingly, both man and beast were vulnerable to injuries, ill health, and other hazards associated with traveling long distances, often in very difficult circumstances.[1] Most animals lost condition and some even died as they made their procession southward to be fattened and sold at London's markets (Thomson 2005). As cattle dealers often supplied animals to and purchased animals from drovers, this meant they played a powerful stockbroker role in the nascent animal-exchange and marketing system. Thus, for the duration of trysts and fairs, a lot of cattle transactions were conducted via dealers.[2]

However, come the nineteenth century the notion of a cattle salesman was beginning to emerge. This, in time, laid the professional foundation for a new breed of selling experts: auctioneers (Haldane 2002). The Napoleonic Wars fueled demand for food, which brought new wealth and opportunities to farmers in Scotland. Some invested in improving their land. Others changed track to become merchants. And a few who became dealers went on to set up livestock markets (Thomson 2005). With increasing industrialization, the old droving system was significantly undermined in two main ways. First, the growth of towns and the enclosure of land thwarted many of the age-old droving routes and removed drovers' access to vital grazing pastures. Second, new forms of transportation—steamboats in the 1820s and railways in the last quarter of the century—provided more efficient alternatives to the long and arduous cattle trails (Perren 1978). Whereas it once took forty days to traipse cattle to London, it now took forty hours by rail. The trade responded by building markets near shipping ports and rail routes (Carlyle 1975; Thomson 2005). In all, these developments were sounding the death knell of the drover's trade and the dealers' monopolization of power. Their demise was further hastened by the eradication in 1845 of a tax on the sale of portable goods off farms (Graham 1999, 46). Although some local dealers tried to jeopardize these sales, by the close of the nineteenth century, "there were 900 livestock markets operating in Great Britain, with auction sales taking place side-by-side with sales by private treaty' (Graham 1999, 45).[3] Gradually, as the buying and selling of livestock began to be mediated by a third party (i.e., an auctioneer), the auctioneer took responsibility for "selling, advertising and auctioning stock to gain a higher price for the seller in return for commission" (Graham 1999, 45).

Agricultural auctioneering firms were typically family affairs: Sons and grandsons followed in their fathers' and grandfathers' footsteps. Such novices were expected to serve out their apprenticeship by starting at the bottom as office boys and drovers and work their way up (Thomson 2005, 61). Ralph Cassidy (1967, 102) lists various qualities valued in the auctioneer: "a good voice, a pleasing personality, good character, the ability to hold and interest an audience, confidence in himself, the ability to judge the value of the items being auctioned, honesty, poise, enthusiasm, and vitality." It would be naive to assume that auctioneers consistently demonstrated high levels of integrity, trustworthiness, and sobriety. Those who deviated from these highly valued attributes undoubtedly would have played a part in bringing the trade and fledgling profession into varying degrees of disrepute (Thomson 2005; Walton 1984). In addition to these desirable personal qualities, a good auctioneer had to be able to gauge the value of what he was selling. In theory, the mechanics of the auction require a number of interested and competing buyers to place ascending bids until only one buyer remains; this contributes to getting the best market price on the day.[4] The auctioneer needs to open the bidding neither too high nor too low because this might discourage interested buyers from participating in the sale. He also needs to decide when to cajole bidders, to squeeze out the final last bid, and when to stop. In typical fashion, this emerging occupational group founded a professional institute in 1886 to legitimate its valuation credentials; this not only enhanced auctioneers' profile but justified their fees (Walton 1984, 26).

The Rise of the "Butcher's Beast" and Having a "Good Eye for a Beast"

Before the 1750s, "the qualities for which animals were valued were not propensity to fatten or early maturity, but their milking capacity or their power of draught. The pail and the plough set the standard; the butcher was ignored" (Prothero 1912, 179). By the mid-1850s, demand for meat from a burgeoning laboring and urbanized populace was growing. In hindsight, this exigency proved to be a pivotal turning point in British livestock farming because animals started to be valued for their capacity to produce meat. An astute English farmer, Robert Bakewell (1725–1795), seemed to be in tune with the changing times and anticipated the farming implications of this mass dietary transition. By the age of twenty he had already tried out livestock-breeding trials on Dishley Grange, his father's tenanted farm in Leicestershire, and within fifteen years he had proceeded to single-handedly run the farm.[5] In many ways, Bakewell probably embodied many of the stereotypical features of an English yeoman as depicted on Staffordshire pottery jugs: "a tall, broad-shouldered, stout man of brown-red complexion, clad in a loose brown coat, scarlet waistcoat, leather breeches, and top-boots" (Prothero 1912, 184).[6] Over time, he refined the process of breeding pedigree stock.[7] By drawing on the experiences of racehorse breeders and applying them to farm animals, he selected animals that he believed best met the butchers' expectations and requirements (Trow-Smith 1959). Up to this point, breeding by non-elite farmers was fairly chaotic: "the haphazard union of nobody's son with everybody's daughter" (Prothero 1912, 181). By adopting the principle of in-and-in breeding, Bakewell bred with animals derived not only from the same indigenous breed and bloodline but also from the same family. For instance, "His bull Twopenny was not only mated with his own mother but was also bred to their daughter, while the product of that mating was bred to his half-sister—again by Twopenny" (Symon 1959, 324). This hastened Bakewell's ability to more readily obtain the fattening attributes he was seeking to elicit from his animals. He successfully applied these principles to Leicester sheep and is credited with developing the first modern breed of sheep: the "New Leicesters" (Franklin 2007, 102).[8] Unimproved sheep took about four years to fatten for market; New Leicesters were ready in half that time. His endeavor to improve Craven Longhorns, a local breed of cattle, was successful but less well received. Although the "New Longhorns" fattened more quickly, this was achieved at the expense of diminishing milk yields. In an area where farmers also relied on their cattle for dairy production, this effect was not so readily appreciated, and for others it cast doubt on the improved animals' ability to reproduce (Prothero 1912; Trow-Smith 1959).

In spite of this, Bakewell's contribution was timely and significant. For the first time livestock were bred for and produced from the butcher's point of view—hence, the significance of the phrase the "butcher's beast." His pioneering breeding technique was also beyond compare, although some contend that it was "his esoteric showmanship and the publicity given to him by the great agricultural reporters of his day which consolidated his supremacy over other contemporary improvers whose real achievements were as great, or nearly as great, as his" (Trow-Smith 1959, 46). Even so, he inspired contemporaries, and future generations of elite and non-elite farmers, in Britain and beyond to try to emulate his success. Bakewell may have generously entertained the many privileged guests who visited his farm, but he remained tight-lipped about the nuts and bolts underpinning his approach. Perhaps unsurprisingly, he blatantly milked his technical know-how for all it was worth. In less than two decades, he could command extortionate prices for selling and leasing his improved rams; this ensured a very healthy return. For instance, "Between 1770 and 1789 saw an increase from five to thirty guineas for a season's use to as much as three thousand guineas" (Franklin 2007, 103). Given all this, Harriet Ritvo (1987, 66) suggests that Bakewell "epitomized the new agricultural technologists" who were coming to the fore and, as such, was the carrier of an alternative stock-breeding discourse to that of elite pedigree breeders.

Gentleman farmers had the time and financial means to invest in pedigree livestock breeding. Those who participated in and presided over the growing number of annual agricultural shows and societies also relished the symbolic opportunities provided by these events to proudly display their prized corpulent livestock. In effect, these public gatherings "functioned as ceremonial reenactments of the traditional rural order. . . . Their rhetoric of service reminded ordinary farmers that the men who could afford to raise prize animals were the natural leaders, at the same time the opulence of the display underlined the exclusiveness of high stock breeding" (Ritvo 1987, 52).[9] Although non-elite farmers were excluded financially from high-end livestock breeding, they were often employed by elite breeders. In effect, ordinary farmers were the brains and practical technicians who, behind the scenes, possessed the necessary skills, knowledge, and experience to create such beasts. Even though both sets of breeders produced fat livestock, they were inspired and motivated by different rationales. While elite breeders courted the prestige, symbolic status, and aesthetic qualities associated with their animals, technical breeders like Bakewell focused on refining and enhancing the utility and edibility of livestock to secure the financial and material rewards from doing so. To all intents and purposes, pedigree livestock began to transcend their everyday realities of the farmyard to become "embodiments of beauty and elegance" (Ritvo 1987, 56). Wealthy breeders and owners commissioned portraits, paintings, and exhibitions of their prizewinning cattle, which visually captured and widely disseminated the desired physical characteristics of the newly improved breeds. For instance, the "Durham Ox," a massive, fleshy Shorthorn improved by Charles Colling, the owner of a small farm near Darlington, sparked the imagination of people around Britain. Bought by a prosperous aristocrat, the ox traveled around England and Scotland for six years in a purposely built carriage: "Not only did people pay to see this beast, but in 1802 alone, more than 2000 people purchased a print of the squarish, roan ox" (Quinn 1993, 150). Despite the unstable and contested nature of idealized breed standards, these pictorial representations influenced the practices of other breeders, who tried to emulate the picture-perfect animals (Quinn 1993). But the fascination with aesthetic attributes generated tension between privileged breeders and the bulk of ordinary farmers, who considered such cattle of little practical use (Holloway 2005, 885).

As dedicated breeders and feeders strove to create ideal breeding and butcher's beasts, they introduced an element of internal competition into the developing livestock industry. The advent of the live auction sale became an important arena for valuing and evaluating livestock. This socioeconomic institution not only gave producers of livestock the public forum to parade their animals and the quality of their stockmanship skills; it also opened these skills to critical appraisal by their peers: "The mart—the weekly auction mart—was the Scottish farmer's stock exchange, not only literally so but in the starker reality of his financial resonance. It was here, as in his fields, that he assessed the performance of his livestock and their future profitability; it was here that his own farming skills were tested and sometimes found wanting by critical peers" (Cameron 1995, 167).

The impetus and expectation for producers to refine and improve their breeding and finishing skills led the mart to introduce special shows and sales of livestock throughout the farming year; this in turn "heightened the competition and punctuated its calendar" (Cameron 1995, 173). Thus, the quality of livestock produced, shown, and sold became synonymous with a *farmtoun*'s reputation. The rather fickle but nonetheless extremely influential effect of buying and selling *nowte* (cattle) at the mart is well illustrated in the following quotation:

> Even if the old men of the farmtouns might not have acknowledged it so, all, intuitively, knew it. From the mart a man might go home like a king: with a lift of the heart at the prices he had heard round the ring and some certainty at least about his immediate future. Sometimes too the mart sent him home in despair to bury his dreams and maybe with the knell of the roup's [auction's] unforgiving gavel already ringing in his ear. It was all a harsh and hazardous business that only the frequent sedation of drams could make anything like bearable. And when it was otherwise, what better cause for celebration? (Cameron 1995, 167)

Moreover, the quality of a butcher's beast is thought to be gleaned from studying its corporal conformation: the shape of its physique, its bodily structure, and the dispersal of flesh over its bones (Holloway 2005; Trow-Smith 1959). This meant that farmers, buyers, sellers, butchers, and judges of cattle had to develop a "good eye for a beast." Such "skilled vision" and visual apprenticeship were acquired and honed through years of practice (Grasseni 2004, 53, and 2005). As some occupational groups were more proficient than others at visually assessing cattle, there was money to be lost by having a poor eye for a beast. For instance, auctioneers, dealers, and butchers developed a particularly discerning visual appreciation of cattle; quite simply, to do their job well required them to gauge accurately the economic potential of the animals they routinely bought and sold. Although farmers and stockmen were skilled at judging their cattle's potential, when it came to the farmer judging the weight of fatstock and matching his judgement against that of the butcher and professional dealer, the farmer was at a disadvantage. This was inevitable. The dealer or the butcher spent his life in estimating the weight of stock. His eye and judgement were his stock in trade and by long experience they became very efficient. The average farmer, however, attended a livestock market only on those occasions when he had stock to sell, which were comparatively few in relation to the number of attendances put in by the professional traders. (Perren 1978, 141)

This rather loose approach to visually assessing cattle was to be tightened. In 1887, a royal commission—the British Parliament's vehicle for investigating problems and proposing new policy—considered complaints that markets were benefiting their owners more than those who used the facilities (Graham 1999, 46). The result was the passing of the Markets and Fairs (Weighing of Cattle) Act in 1887.[10] Although it was mandatory for all cattle markets in Britain to install weighing machines or weighbridges, it was not compulsory for market owners and sellers to use them. Over time, weighing was slowly, reluctantly, and inconsistently implemented:

[The weighing machines] were unpopular both with market owners, compelled to incur an expense for which they saw no need, and with farmers, who were accustomed to purchase and sell animals by judgement of eye rather than by weight. . . . Market authorities did only what they had to in order to comply with the letter of the law; weighing machines were unattended, often inaccurate, and generally inadequate for the requirements of the market. (Perren 1978, 140)

Farmers were frequently "out of pocket" due to their lesser ability to judge the live weight of animals. In 1893, the agricultural journalist Westley Richards worked out that the typical British farmer was forfeiting 45 shillings on every fat animal sold.[11] For the beef sector as a whole, this generated an annual loss of up to £5 million. Given this, he said, "A weighing machine is as necessary an implement on a stock farm as a plough on an arable farm if any accurate knowledge of the size, growth and value of stock is to be obtained" (quoted in Perren 1978, 142).

By 1912, fatstock farmers in Scotland, unlike some of their farming colleagues in England, had embraced these legislative changes. In agricultural regions of Britain where dairy farming was prevalent, there was less concern about the weight of "dried off" dairy cows that were fattened for beef. As these animals were not primarily bred for meat, their carcasses were usually worth less. In contrast, farmers who specialized in finishing beef cattle had more of an incentive to reduce any discrepancies in the live and dead weight of the animals they sold—hence, their increased use of weighing machines at markets.[12] This difference between farmers also alerts us to the increasing specialization that was occurring within this sector:

> Breeds of cattle in Britain in the first half of the nineteenth century were, therefore, beginning to exhibit their modern division into the three groups of beef, dual-purpose and milch beast. Before this period all the breeds . . . were truly triple-purpose stock; the bullocks worked and beefed, the females milked, worked and beefed. Pedigree fancy, allied to the demand for meat, then stimulated the evolution of a specialist beef type. The market for liquid milk next made the development of a specialist dairy type profitable. The marriage of the horse to the light iron plow released both types, beef and milch, from the bondage of the yoke. (Trow-Smith 1959, 265)

Livestock Go to Market: Valuing Breeding, Store, and Prime Animals

In keeping with early locations of other major livestock markets in the United Kingdom, the main mart in northeast Scotland, was adjacent to a railhead. One auctioneer recalls the problems of droving livestock through the city streets en route to the station: "There was always a beast landing in somebody's garden, or a car mirror got knocked off, or a bashed wing. . . . [T]he cow pats over the pavement. The people used to know what day it was happening and what time; they used to be outside with their brooms." He also remembered the day a beast escaped from the mart and it made its way into the city center, causing a furor: "There was a lady with a pram outside a [department store]. It went for this lady *wi* a pram and bashed a window, and after that the Health and Safety [officials] came up to . . . [the] mart and they just filled a book with things that were wrong with the mart that had to be improved. It would cost a million pounds to do the mart up to their standing." This led to the 1990 relocation of the mart, founded in 1870, to a field site with good road links. In some ways, this was a sign of the times: Heightened urban sensibilities about the production and slaughter of livestock, increased awareness of workers' health and safety, as well as the need for better animal-welfare and lairage facilities all contributed to the shift. Since 1990, the mart has carried out its business from this site and is reputed to be one of the most modern livestock auction markets in Europe. From the outside, the huge metal-clad building looks like a shopping mall. Inside the split-level center, it is bright and roomy. It caters to its customers' needs with toilets and eateries. One food outlet offers modern fast food, such as burgers and chips, while the other serves traditional rural fare; many of its meals are based on beef. The center also houses a number of farming-related services, such as financial advisers, animal-feed suppliers, and animal-health specialists, and representatives from a range of agricultural associations, including the National Farmers' Union of Scotland and the Scottish Agricultural College. Providing access to such services has its roots in early major livestock markets, so this is not a new feature (Carlyle 1975, 450). In the foyer, there are customer notice boards and an arena where seats can be gathered to auction small-scale items, such as eggs and vegetables. The mart has three sale rings, a sizeable hall to exhibit livestock, and the capacity to accommodate 17,000 animals. It also has lairage facilities, which provide respite for livestock that have traveled from afar.

Each ring is semicircular, heated, covered with sawdust, and surrounded by safety rails and red plastic seats. The seats immediately flanking the sale ring enable serious buyers to get a good look at the livestock paraded in the ring and allow them to catch the eye of the auctioneer should they decide to place a bid. Similar to the layout of a large lecture hall, rows of seats ascend upward away from the ring; this tends to accommodate family members of buyers and interested spectators who can watch the sale from a distance. On each side of the auctioneer's rostrum are manually operated gates that guide animals in and out of the ring. The men responsible for opening and closing the gates work in the penning area immediately behind the sale ring and wear gray overalls. Auctioneers wear white coats during the sale, and livestock managers in the penning area wear blue coats. This color-coding of overalls depicts the hierarchical differentiation of auction staff in which lower-level workers are coded gray; middle-ranked workers, or "gaffers," are dressed in blue; and auctioneers draw on the symbolic power of the white coat, which is associated with high-status professional "experts" such as doctors and scientists.[13] In front of the auctioneer's rostrum is a metal shield that ring men can slip behind should they need to protect themselves from unpredictable animals that might charge.

Behind the sale ring are what handlers colloquially call the "*byre* (cow-shed)" and penning areas; this backstage area is fed by twenty-one loading bays. Livestock haulage drivers and farmers deposit and pick up their live cargo from this point before and after the auction sales. This leads into an intricate and vast sea of metal pens and passageways that allow groups of animals to be moved efficiently from the bays to the auction rings and back again. The massive extent of this penning area can be fully appreciated from an inspection gangway, or "catwalk," suspended directly above, a vantage point that provides members of the public, visitors, and buyers with a view over the entire area without risking any harm from the animals.

For example, during my time at the mart, a member of staff conducted public tours of the mart premises. Four tours were scheduled per week, each of which was attended by up to fifty people. The tours tended to attract groups of children, teenagers, and couples from the local area. The mart was especially keen to target tourists and people who were not from an agricultural background, as one of the key aims of the tours was to better inform and enhance non-farming groups' understanding of what goes on at the mart. After two weeks of attending the mart, I recorded in my field notes:

> The noise level is constant. You have cattle and sheep mooing and baaing, respectively. You have the noise [at] the loading bays [where] the livestock transport lorries draw up and open their gates and doors; the pen gates [are] banging; workers [are] shouting at each other—pen numbers—or to do something like "race" or perhaps just shouting. They shout at the animals to encourage them to move; some whistle, some bang the pens, some flail a plastic flapper, some speak to them, others gently touch them, some run at them making a whaa noise. Some just walk quietly behind the animals. Gates are always opening and closing. Then you have the noise during the sale: the auctioneer and the head manager conveying via the microphone system who has bought the animals so that the marker can mark the animals accordingly. You also hear the crush crate doors being operated with the cattle at the other side.[14] Then there is the smell of the place. One is aware of urine and feces; the smell seems to penetrate deep into your clothes [and stays] with you even once you are home. . . . There is an incessant noise level—it provokes a sensory overload. . . . The bareness of the place, the concrete floors, cold air [and] metal gates makes the work environment [in the byre] an austere place.[15]

To illustrate the processing of animals from the loading bays via the sale ring and back to the holding pens, I use the example of sheep. As mart workers receive sheep at the loading bays, they book them in. This involves checking accompanying paperwork to confirm who owns them and recording the number of animals deposited. This piece of paper goes with the "lot," which refers to the flock before it is separated for marketing. The sheep then go through the "race" and "shedder." The race is a narrow, metal-sided channel with a gate that is manually operated by workers to skillfully and speedily "shed," or separate, a single stream of sheep according to their breed and perceived weight. This process is very efficient as long as the handlers work as a team. If someone forgets to open or close a gate at the right time, it can cause mayhem, confusion, and frustration. Separated animals are graded before they are weighed. When the time comes to sell breeding and prime sheep, the sale ring is arranged accordingly. When batches of prime sheep enter the ring, buyers have to be able to physically handle their backs and tails to assess the quality of the animals and thus estimate their value. This is not required with breeding sheep. Unlike fat cattle, prime sheep are not so easily visually assessed. The "marker" is a person who sprays pre-coded color combinations of paint onto the fleece of newly bought sheep as they leave the sale ring; the markings depict the main buyers and dealers. There are about twelve different paint cans, and the person responsible for this task must be able to translate the named buyer into his or her idiosyncratic color-coded symbol. Marked animals are then ushered into holding pens to be collected by their new owners later that day. This system helps mart workers process large volumes of animals as efficiently and accurately as possible.

Charles Smith (1989, x) suggests that "auctions serve as rites of passage for objects shrouded in ambiguity and uncertainty [and that] most auctions can be classified into three basic types: commodity-exchange auctions, collectable-dealer auctions, and one-of-a-kind-sale auctions." Clearly, each of these auction types would be governed by "its own communal structures, rules, and practices, which determine everything from who can participate and the form of bidding to relationships among and between buyers and sellers and the role of the auctioneer. Despite these diverse forms and practices, all auctions serve to establish socially legitimated definitions of value and provenance" (Smith 1989, x). Livestock auctions are an example of the commodity-exchange classification. Cattle and sheep are relatively ambiguous commodities that have no "standard" value,[16] which contributes to the difficulty of assigning a stable price to them; such circumstances endorse the use of auction systems (Cassidy 1967, 20).[17] As one senior auctioneer observed, "Farmers have a really . . . rough time of it, because they *hivna* [have not] a clue what they're going to get for stuff when they buy it or when they start breeding it, and that's a way of life, really." Attempts to discover or establish the economic value of livestock lie at the very heart of this method of selling. Although the pricing of livestock cannot be isolated from national and international trends of demand and changes in the costs of production, it is also important to consider how more local elements of the auctioning process, such as the role of the auctioneer in "price discovery" and characteristics of the animate products being sold, contribute to valuing livestock.

One way to increase our understanding of how food animals come to acquire their monetary value and commodity status is to consider the pivotal role auctioneers play in the process. Most auctioneers are farmers' sons (occasionally daughters), and their early socialization into farming and animal husbandry means they enter their chosen profession with varying degrees of insider knowledge about the livestock industry. For example, some of my auctioneering contacts had been born into mixed and dairy farms, while others had come from crofting and shepherding backgrounds. The older generation of auctioneers recounted a long and continuous record of working in the mart, having typically served their time by starting as office clerks. One senior auctioneer explained that it was fairly customary for farmers to have a word with managers at the marts to see whether their sons could be taken on as clerks:

> The mart policy at that time was to take on youngsters—fifteen years old—that left school and train them in the way that they felt was correct for being a useful person in the Company. . . . Then you moved on to being involved in a rostrum doing some clerical work, and then you moved up through the next step, . . . maybe making up accounts or doing sales until in fact you became senior clerk and were doing book work and you got a vast experience of the whole thing, all the way through. Of course, you were really involved with the stock; each day you were out in the pens and booking stock. So you were amongst them all the time and got terrific, terrific experience in stock handling and what stock was all about. And of course meeting people. . . . [Y]ou were either getting stock from them or trying [to] sell stock to them. You just met everybody, which was a great basis for your future as an auctioneer, if indeed that's [where] you wanted to go.

Nowadays, trainee auctioneers are expected to have academic qualifications in addition to an aptitude for working with livestock. A newly qualified auctioneer summarized the range of subjects he had to study to gain his professional qualifications: surveying; tax and accountancy; forestry; law of landlord and tenant, which covers valuations of incoming and outgoing tenants; law of contract sale of goods; and agricultural valuations in Scotland. The link to real estate is not new. In fact, many of the first auctioneering firms diversified from real estate into marketing livestock. The connection continues today, as many livestock auctioneers are also chartered surveyors and estate agents (Graham 1999, 46). The auctioneer also explained that junior auctioneers tend to develop their skills by selling items of lower value, such as eggs, poultry, calves, and sheep. His remark was a fine example of the livestock hierarchy. Farm workers often talk about "cattle, sheep and pigs," the order representing descending monetary value.[18] Another auctioneer noted that, because one cow is worth at least twenty lambs, from "a farmer's point of view it would be higher status to look after a £600–£700 animal compared to a £30 lamb." Of course, notions of value may be regionally mediated, too. For instance, Aberdeenshire is especially associated with finishing store animals for beef-cattle production, so this probably influences the high status of cattle in the region. The Scottish Borders, Caithness, and Shetland are high-density sheep-producing regions, which may enhance how sheep are perceived and valued in these locales (Thomson 2008, 2). However, irrespective of this caveat, commercial cattle will always be worth more financially than sheep and pigs. Unsurprisingly, those who work with these economically ranked animals are equally appraised and ranked according to the species with which they work. This is illustrated by a telling comment the newly qualified auctioneer received from a farmer: "When are you going to move on from sheep to become a real auctioneer?"

Although auctioneers have to have a good eye for a beast and strong animal-handling skills, they also need to get along well with people and have good impression-management skills. As a senior auctioneer explained:

> You've got to get on with people and undoubtedly you've to work hard because it *disna* [doesn't] just happen. You advertise a sale, and we're going to sell whatever next week. . . . [Y]ou get your entries, and they come, and that's it. Sounds awfully easy, but it's *nae* like that. You've got to be out there talking to folk. . . . [Y]ou've to build up a reputation. You've to build up a situation where folk know that you're going to [do] a decent job; they've got to *hae* confidence in the person who's in there doing it, and you can only do that from starting way down here and speaking to folk all the way through. . . . So when they book cattle, they're confident that you're going *tae* have done sufficient work to get buyers there, and you're going *tae* have a sale that's meaningful.

Other auctioneers suggested that one has to "look the part" and "look confident." Clients also appreciate the "personal touch," so it is very important to remember their names and relevant details shared in previous conversations. For others, it was crucial to be seen as "honest and trustworthy." Overall, auctioneers appeared to be acutely aware of, and highly reflective about, their role in ensuring the sales they preside over are regarded by buyers and sellers as honest and credible. In other words, the auctioneer, "who in nearly all instances is the star of the show, remains the heart of any auction" (Smith 1989, 116). But how exactly does the auctioneer contribute to establishing the economic worth of the livestock animals he or she markets?

A senior auctioneer provided a way into this issue when he outlined the aims and objectives in his company's mission statement:

> Our job is to get a value, to fix a value, *nae* fix a value but reach a value, the proper value of what an article's worth. The only real way to find it is by auction. You can put up something and say it's worth £500, but if nobody wants it, it's *nae* worth £500. . . . So we've *tae* get clarity and find where its real value is to people who are going to sell that day. So our job is to get buyers and sellers to meet so that we can get a fair price and a good average price for that article and market it, obviously, to the best of our ability.

Another auctioneer described his job as one of "price discovery" because his role was to ascertain how much a buyer would pay for animals on sale. In effect, he said, "We are the broker, . . . and it's just a price discovery system trying to establish what an animal or animals are worth in terms of hard cash. And, of course, this is the ultimate aim of all commercial businesses: to realize as much cash as they possibly can." When asked to explain in more detail the actual processes involved in realizing this goal of "price discovery," he initially responded, "That's a thousand-dollar question!" then continued:

> Everything has a value to someone. The ultimate aim is for the buyer when he looks at some cattle he considers that he would give X for them because he'll have to keep them for a period of Y and he's going to give Z for them when he puts them away to the abattoir. So he is prepared to give whatever level of money he's prepared to give for them, as is many others. . . . [W]hat makes the way you arrive at the price discovery system is by the amount of people that you have attending the auction sale, and they are prepared to bid up to the particular prices. That's our job, to collate these bids and get [the price], as far as the seller is concerned, as high as we . . . possibly [can]. But . . . all prices are driven by the end product price—that is, the price that the supermarket will pay for beef—and that, again, I suspect will be driven by the price that the supermarket can [charge to] sell it to the housewife or the consumer.

These accounts appear to assume that buyers and sellers are essentially independent, rational actors actively pursuing their economic interests and that auctions are "socially uncontaminated mechanisms for matching the individual preferences of buyers and sellers. The goal is exchange, with the determination of the selling price as the means" (Smith 1989, 162). But as Smith notes in his sociological analysis of different auction systems, "The practices, places, participants and conventions of the auction itself—what could be called its 'social structure'"—also play a significant part in shaping and contributing to the marketing process (Smith 1989, 162).

A key feature of the price-bargaining process is that it becomes easier to determine the value of cattle as they approach the endpoint of the production process (i.e., prime animals slaughtered for meat). This is so because professional buyers such as butchers assess and price prime cattle in terms of their perceived conformation and fatness as set out in EUROP, a carcass-grading system used by abattoirs.[19] These animals are sold within a relatively tight price margin, so there is little variation between animals. As their selling price is hugely determined by their physical weight, their economic value is fairly predictable. In practice, this means that, once representatives of the main slaughterhouses have set the price with their first few purchases, prime animals are processed very quickly through the sale ring. Breeding animals, by contrast, have an indeterminate floating value, which may be influenced by a variety of factors, such as their breeding history, age, looks, and reputation of the selling farm. Thus, their price is more variable, and the selling of these animals takes more time. In addition, breeding animals are bought by a large number of farmers, while prime animals are bought by a small number of professional buyers. As a result, it would seem that the career path of the food animal (i.e., breeding, store, or prime) combined with who is purchasing the animal are also important in terms of how livestock are marketed and valued.

For example, while observing sales, I noticed that breeding animals spent the longest period of time in the sale ring and that auctioneers provided a fulsome history of each animal, summarizing its potential. Store animals were processed faster and in small batches with less background information. Finally, prime animals were virtually in and out of the ring, with little more than their weight being mentioned. I explored the reasons for these differences in marketing with two senior auctioneers, who noted that in the prime-sale ring, "You've got specialist men round the ringside who know that if this bullock comes in, it's worth 105 to 106 pence [per kilogram]. An auctioneer, if he knows his job, knows it's worth 105 to 106, so he can start it at 100 and 1, 2, 3, so you know it's going to make 104 to 105 or 106, and that's the end of the story." The "specialist men" doing the buying are regarded as "professional buyers" and are usually dealers and butchers. Because these buyers "are in a market probably facing an auctioneer three, four, or five days a week, . . . they know exactly what they can give for them because they are part of the chain that's closest to the wholesale: the supermarket or the butcher. . . . [T]hese people just need one quick look at a beast and make up their mind what they're going to give."

Store-cattle buyers, who are typically farmers and dealers, require extra time to assess the potential of such animals and will need to carry out quick calculations to determine how much they are prepared to bid. Once these animals have been fattened up, they go on to supply the professional buyers in the prime ring. A cattleman explained how butchers and, increasingly, supermarkets indirectly set the price level for store animals: "Whatever the butchers are paying for animals that are being slaughtered, that will set the price for store cattle that are going through the ring. . . . If it's a better quality animal, then it'll get a higher price, or not such a good animal won't get such a good price. . . . But it really does go down to what it's worth at the end of the day to the butcher. That's what the farmer will set his price levels [on]." The price-setting function of the mart ring ripples outward. Even though some farmers might prefer to bypass the mart's commission and arrange private deals, or sell directly to abattoirs, in agreeing their price they still take account of recent mart prices. As one auctioneer said, "Lots of farmers who feed cattle or sheep maybe put them all to the abattoir. But they're still watching the mart report every day. They're always guided by the mart. If there's no mart going on, they've lost it; they won't know if they're getting the right price for their stock or not." It seems that the notion of the "butcher's beast" remains relevant in contemporary livestock-auction systems, as it significantly informs pricing practices in these settings. It also quite literally shapes the productive ideal physique of fat cattle. For instance, this cattleman is acutely aware that "quality" refers to shape: "It's all about the shape of the animal." But the shape is based on "what shape the butchers want the animals to be when [they come] to slaughter." The ideal shape is a moving target, but some features never change:

> The overall size of the animal shifts a *wee* [little] bit. Lately, the supermarkets are telling the butchers that they're looking for smaller animals, but that's only to a point. . . . [Y]ou need the rounded hips and decent rump. That hasn't changed; you still need that muscle to be as big as possible. But you can get a quality small animal or quality large animal; it hasn't really changed that much. They don't like them quite so big, but they still have to have shape.

The economic valuation of livestock is also determined partly by the sex of the animal and the level of financial subsidy associated with store animals. This is illustrated by an account from a female commercial farmer who deals with the paperwork associated with processing cattle subsidies. She suggested that female animals are "always going to be worth less in the commercial ring than a male for fattening, but not necessarily for breeding. And, of course, a male animal is more valuable these days depending on the color of its paperwork: the CCDs [cattle control documents]."[20] She went on to explain how those documents work:

> A male animal is born or a female animal is born, and the first thing you have to [do] is put an ear tag in both [of its] ears. The ear tag has to correspond in both ears; it has to be identical. Then you register that animal with the . . . British Cattle Movement Service. On that passport—it's a checkbook-style thing these days—it has the dam's number and, if you can be bothered, the sire's number, the date of birth, the sex, the breed, and which farm it was born onto.[21] So the male animals at the moment have paperwork which is valuable, depending upon what color it is. Initially, it starts out as white, and when the animal is eight months old, you can submit that paperwork to the Department, [which] will put the animal on what is known as a "retention period," whereby you have to keep it on the farm for a two-month period and you will be issued a blue piece of paperwork. That should maybe generate £80 to £90, depending upon the national number of cattle that are turned around on an annual basis within the United Kingdom, so it's variable. If . . . 100 cattle [were] turned around in the United Kingdom, I might get £100. But if . . . 200 cattle [were] turned around, I might just get £50. When the animal that is now on a blue CCD becomes twenty-one months old, I'm allowed to turn it around from a blue to a pink, and that will generate another bit of money, depending upon the national number of animals turned around on an annual basis. So value—sex, paperwork, and how good-looking [the animal] is.[22]

One of the bureaucratic implications following the BSE crisis was that cattle had to be traced from "farm to fork."[23] In principle, ear tags and "passports" individuate and document all movements cattle make, from birth to slaughter.[24] But these paper trails and subsidy applications have created a lot more administration for those involved in producing, dealing, marketing, and slaughtering cattle. Members of the older farming generation, who tend to be less computer-literate, are finding these developments quite a burden. In their day, a farm estate manager said, paperwork was "frowned upon"; farmers often kept the information they needed about their animals "in their head." Their wives or female partners usually oversaw that side of the business, but today many spouses work outside the farm to supplement the family's income. A commercial farmer who has 300 breeding cows and finishes their progeny now may have to do the paperwork himself. Keeping the paperwork up to date during calving time is especially difficult, as legally the farmer has to tag all the calves within thirty days of birth.[25] This has since been reduced to twenty days. "Everything's getting tighter and tighter," the commercial farmer said, "and that means you've got to do all the paperwork in the middle of calving." This means finding the time "to come into the office, to sit down, and start applying for passports for all these calves, just at the busiest time o year." But, he also said, "[Farmers] can't afford to be wrong. We are not allowed to make mistakes. . . . There is not such a thing as 'I'm sorry.' Everybody else is allowed to make a mistake—the Department, the British Cattle Movement Society—but when it comes to us, no. The thing has to be spot on, absolutely spot on. So I feel that if I'm involved, I know the calves, the numbers, what's expected, so it's easier if I do it myself."

If forms are completed incorrectly or application deadlines are missed, farmers stand to lose a lot of revenue. These points were echoed by the female cattle farmer. She said that until ten years ago, her husband was the most important person on the farm, "but now it's the person that fills in the right box, at the right time that generates the most money. And I would say the balance of things has swung in my direction." She used a hypothetical example to show how important it has become for farmers to have the correct paperwork: "If there was a choice between a calf lying out there in the field and needing treatment or me filling in my Beef Special Premium form, it would be far more financially advantageous for me to sit in the office and watch that calf die and get the dashed paperwork correct, which goes against the grain of most livestock people." However, she believes that it has been easier for her age group to come to terms with the fact that "the paperwork is more important than the livestock. . . . For an older person, that's very difficult to do." Thus, the younger generation seems better equipped to deal with these bureaucratic changes, and some of the large farm estates with more resources have responded to the new pressures by employing full-time secretarial staff.

The breeding sales are particularly distinctive. As I have noted, these sales are slower, and the auctioneer provides far more background information about these animals. Farmers are the main buyers of breeding animals, and it is important to note that unlike prime animals, which are slaughtered by the time they reach thirty months of age, breeding animals can remain on the same farm for ten years or more, as long as they remain productive. Thus, the purchase of breeding animals is a significant and substantial financial investment. Unsurprisingly, said a senior auctioneer, buyers of these animals "wish to study exactly what they're going to bid for even before they start bidding for it to establish whether they think it's exactly what is needed and suitable for their farm. So the whole thing takes that little bit of time, and people just need that few seconds more before they are encouraged to start offering for it." There is a lot of variation among breeders; thus, different buyers look for particular characteristics in an animal so that it best serves the needs of their farms. For example, commercial breeders often favor Limousins, Simmentals, Charolais, or Aberdeen Angus, with their breed preferences partly informed by "aesthetic taste, sentimental attachments and personal history" (Grasseni 2004, 45). Some look for calves born in the spring or the fall. A senior auctioneer noted that "it takes folk a lot longer to make up their mind about an animal: what its worth is, if it's going to fit in *tae* their program—it's a slower thought process." Drawing on his many years of conducting breeding sales, he explained:

> You maybe think when it comes in [to the sale ring] that this is a unit worth about £900—a cow and a bull calf.[26] You get started eventually about £700. You just think it's worth about £900, there's *nae* fixed, it's *nae* like 104 to 106 pence per kilo; it's 'round about there. It could make on a bad day £820; on a good day, [with] a couple of real keen guys, you could get £1,000 for it. So there's . . . nobody [who will] fix the price *o* a breeding animal to its last £10. You can put in what your thoughts are on it and what they [are] likely to average, but you know it's depending on who's interested. So you've got to give them time. You've got to tell them the story about what's all in there, and quite often, you know, you'll start a beast at £800. . . . It's a lot better to say, "Look at this calf: This was born in April; it's in calf again." All you're doing is filling in time to give them time to think if they're going to bid £820 for it, and you're also giving them information. It's to make your oration in the box more interesting to the folk around the ringside. There's nothing worse than somebody just prattling on to a price all the time, particularly if it's *nae* easy understanding what they're saying. . . . You've got to say something interesting. In fact if *ye canna* say something interesting, you *sudna* [shouldn't] be there at all.[27]

Thus, different sales have different rhythms. Prime sales have a fast and steady rhythm, while breeding sales are much slower and irregular. Anyone who has ever attended an auction is well aware of the auctioneer's chant. Outsiders unfamiliar with this verbal auction ritual might find it difficult to decode; to those in the know, however, the chant provides signposts to where they are in the auctioning process (Smith 1989, 117). As the senior auctioneer noted, auctioneers "fill in time" by inserting pieces of information to help potential buyers make up their minds. Thus, the chant can be interspersed with "fillers." Given the increased uncertainty surrounding the valuation of breeding animals, as they are further away from the end product and their weight is fairly irrelevant, the chant provides much needed structure and meaning where little exists. As Charles Smith (1989, 117) notes, "The key to a good chant is not only to monitor the bidding but also to establish the cadence of the bidding. It does this by controlling the size of bid increases as well as their timing. It manages to take what is a very erratic, disjointed process and meld it into an ongoing, comparatively harmonious process. Like any music, it provides a unifying rhythm or theme."

What is common to all of these sales is that livestock are marketed through the sale ring while alive. However, since the beginning of the 1990s, alternative methods of selling livestock have come to the fore, such as deadweight marketing systems.[28] It is to this system that I now turn.

Livestock auctions mushroomed during the nineteenth century to become the dominant method of marketing and selling commercial livestock. Since 1945, a substantial number of markets have closed throughout the United Kingdom. In 1940, there were 554 markets in England. During the early 1980s, this number fell to 312, and by 1992, it had fallen again, to 246 (Wright et al. 2002, 477). In the United Kingdom, there were over 800 markets in the 1960s; by 2001, this number had contracted to 170 (Department for Environment, Food, and Rural Affairs 2008, 4). Increasing pressures to streamline markets have been partly linked to the "capital costs of modernising or meeting regulatory requirements, the capital costs of relocating markets, and decline in livestock numbers in surrounding areas" (James Jones and John Steele, cited in Wright et al. 2002, 477). Some obvious implications of these cutbacks are that livestock have to travel longer distances to be sold at the remaining markets. This heightens concerns for the welfare of the animals and exacerbates concerns about the spread of highly infectious diseases.

Such fears were realized in 2001 during a major outbreak in the United Kingdom of foot-and-mouth disease, a "highly contagious viral disease of cloven-footed animals" (Woods 2004, 341). In 1967, the year of the last major outbreak of the disease, 442,285 animals was slaughtered, but the virus was more geographically contained (Department for Environment, Food, and Rural Affairs 2008, 5). In contrast, more than 10 million animals were slaughtered in 2001 (Woods 2004), and it is thought that the restructuring and the increased movement of animals within the industry had greatly contributed to the gravity of that outbreak. Prime sheep can frequently change hands during their productive lives: A hill farmer might sell his lambs to a dealer who might sell them to a lowland farmer to finish them off. The lowland farmer might sell the animals to a dealer who might pass them on to another farmer, a dealer, a market, or an abattoir (Department for Environment, Food, and Rural Affairs 2008, 4). Given this pattern of sheep movements and the airborne transmission of foot-and-mouth disease, an epidemic was in the making. Although the disease was initially identified in infected pigs in an abattoir in Essex, it was thought to have originated on a farm in Northumberland. Delays in locating the primary source of the disease meant that infected sheep had already been sent to Hexam market. From there, dealers moved sheep to a market in Cumbria and from there infected sheep were moved to a number of other markets, including Carlisle, Devon, Dumfries and Galloway, and Cheshire. As the Department for Environment, Food, and Rural Affairs (2008, 2–3) notes, "Because the primary outbreak occurred hundreds of miles away and probably many days before, it had already spread to the sheep marketing network before it was detected." This meant that "infected sheep were crossing the country in hundreds of separate movements, and coming in contact with other livestock" (Department for Environment, Food, and Rural Affairs 2008, 2–3).

Remaining market operators also have had to reduce their staff and diversify into a range of non-farming services to offset the livestock side of their business. Given these structural changes, farmers can now opt to bypass the live sale ring and sell their animals via an alternative marketing process: the deadweight system (Wright et al. 2002, 482).

Farmers in northeast Scotland are no exception. In 1991, the mart set up a deadweight marketing system for lambs that enables farmers to sell directly to abattoirs or butchers. This means that lambs bypass the live ring, and staff try to secure the best market price on behalf of their clients with buyers at abattoirs and butchers throughout the United Kingdom. In theory, some abattoirs specialize in lean, small lambs destined for export to countries such as France or Italy, while fatter and heavier lambs might be purchased by local butchers. So lambs are selected to meet different specifications and priced accordingly. However, in practice, an auctioneer suggested, "Most abattoirs are looking for the same type of sheep. That's the biggest snag." Farmers might decide to use the live or deadweight system depending on the time of the year they are selling animals, the market price of lambs at the time, and the breed of sheep they are selling. Certain commercial breeds of sheep are less popular with butchers than others.[29] As the auctioneer explained, "If you're selling lambs live, you're looking for . . . Suffolk, Cheviot, Texel—tight-coated breeds. The ones that are mostly penalized are Mules, Grey Face, [and] Black Faces—hill-type breeds. But once they're hung up, how can I phrase it, they're all the same; they're just a general commodity *wi* the same average price. . . . [O]nce they're dead, they're *nae* treated as a Grey Face lamb; it's just a carcass—another 2 or another 3 L, or whatever the grade is."

The "grade" refers to the EUROP carcass-classification grid produced by the Meat and Livestock Commission, which has been used in abattoirs throughout the United Kingdom since the 1970s (Graham 1999, 123).[30] There are separate but similar classification systems for cattle and sheep carcasses. The grid has two assessment dimensions: conformation and fatness. Carcasses are appraised according to their shape and level of fat distribution and will be placed into one of five conformation grades and into one of five fat classifications.[31] "Class E describes carcasses of 'outstanding shape,' while at the other extreme P represents 'poorly muscled carcasses of inferior shape.' Fatness ranges from 1–5 whereby 1 refers to 'very lean' and 5 is 'very fat.' Bands 3 and 4 are split into two additional bandings: L for 'leaner' and H for 'fatter'" (Graham 1999, 124).[32] Since 1992, this classification system has been applied to all cattle carcasses throughout Europe, and in 1997 it became mandatory for sheep carcasses. One of the main functions of EUROP is to provide a common baseline for the amount of saleable meat on each carcass for producers, processors, and retailers; the lower the fat banding, the more saleable the meat (Quality Meat Scotland 2004, 4). For example, it is thought that about 70 percent of a beef carcass and about 90 percent of a lamb carcass is saleable meat.[33] The difference in percentages of saleable meat is linked to how much bone is removed or left in each type of carcass. Cattle tend to have most of their bones removed, while some cuts of lamb still have the bone present. Moreover, since the target baselines for cattle and sheep range from R2 to 4L and R2 to R3L, respectively, this has provided a template for a standardized pricing grid for buyers and sellers. Thus, carcasses that fall within the target area will receive the going rate, those at the upper end will earn more, and those falling short will be penalized.

The EUROP system gives livestock producers, stockmen, auctioneers, buyers, and sellers a common framework for anticipating prices. However, they can disagree about the appropriate classification for a live animal. As one auctioneer acknowledged, half the battle is getting farmers to have confidence in your appraisal: "I'll get it 75 percent right, but then you have instances where it'll go to an abattoir, and it's maybe a different man assessing the stock on the hook, and he could be totally wrong, as well. So it knocks the whole thing. . . . [T]here's too many individual[s]; there's me assessing them on the farm for a start, then there's [the Meat and Livestock Commission] assessing them, and there's what the farmer thinks." Some workers appear to be more proficient at visually assessing animals on the hoof, while others find it easier when they are on the hook. Despite attempts by the industry to standardize the assessment, grading, and pricing process, the "visual mode of evaluation prioritizes a subjective appraisal of the [animal] body, and is associated with particular forms of aesthetic judgement undertaken in specific places and settings" (Holloway 2005, 887). There continues to be a division of visual labor whereby workers located throughout the production, marketing, and slaughtering process not only are socialized into different ways of looking but also have varying degrees of experience in terms of augmenting and claiming the accuracy of their "skilled vision." Of course, such disputes and discrepancies are not new. As discussed earlier, farmers lost money to butchers and dealers before weighing machines were introduced in markets. Having a good eye for a beast is thought to enhance finishers' abilities to assess the percentage of saleable meat on prime livestock as they prepare them to become deadstock. This visual aptitude is based on an assumption that they can, indeed, "judge what is inside from the outside" (Holloway 2005, 894). However, with more emphasis being placed on genetic interventions to create better-quality beef animals, the extent to which having a good eye for a beast will continue to be valued in the cattle sector remains to be seen.[34]

Another attraction of the deadweight marketing system is that it saves the farmer from attending the auction. There seems to be a generational component to this change: Younger farmers are less likely than their predecessors to go to market. One auctioneer said that he had noticed that "a lot o the younger ones are more interested in the technical side of [sheep]. . . . Some of them will try and improve their breeding for the following year, change their *tups* [ram], or whatever." In contrast, more "traditional guys" go to the live auction because their fathers did it: "It's tradition." Some farmers seemed to think they should see their animals being sold. This was partially linked to their having "some pride in the stock that they're showing off," whereas those who put their animals deadweight *"arnae sae* [are not so] worried about the pride, just the money at the end of the day." Undoubtedly, the live auction is an opportunity to showcase higher-quality animals and secure higher prices. Auctioneers explain that an above-average beef animal would earn about £100 extra if sold live but only £30 if sold dead. If a local butcher and a "small retail guy" are looking for high-quality animals, as opposed to commercial "middle-of-the-range ones," they might bid against each other, which would raise the selling price. The inverse is also the case. "*Orra* [shabby]" or "plainer animals can be stickier to sell in the ring," said an auctioneer, so the deadweight system might be an alternative way to market poor-quality animals.

In summary, this chapter traces the rise of livestock-auction markets and auctioneering in the nineteenth century and explores why this method of marketing is used to sell livestock. Establishing the value of cattle appears to be closely tied to the purpose for which the cattle are being used. Because there is more variety in the interests of buyers and in the range of characteristics that might appeal, breeding animals are much harder to value. Breeding animals can be individually valued by both sellers and would-be buyers. And a farmer who thinks his or her animals have some characteristic that will attract the attention of bidders will want to present those beasts in the live sale ring so that their favorable qualities are fully appreciated and taken into account. In contrast, prime cattle ready for slaughter have a fairly predictable value. Once consumer demand has helped set the price that butchers and supermarkets are willing to pay, the value of a finished beast can be ascertained from its weight. Thus, the nearer the animal is to being meat, the easier it is to price. Finally, store cattle are more likely to be sold in batches, as are plainer animals that are not expected to impress buyers. Beasts of inferior quality will tend to be sold deadweight, as opposed to going through the sale ring.

What is operating in the auction system is a pattern that seems to run through the entire productive process. Animals that have experienced varying levels of individual attention both before and after auction will tend to be attended to as individuals. Animals that are about to lose their sentience are more likely to be treated as commodities. One productive context that accentuates the individuated status of livestock is hobby farming. It is the nature of hobby-livestock relations that we now consider.

5

"The Good Life"

Hobby Farmers and Rare Breeds of Livestock

Och it's folk that . . . just want to keep a few pets. I mean, they'll never *mak* [make] money at it, there's a few *o* them thought they could make a living out of a few acres. . . . But, I mean, they fail miserably; there's *nae* hope. . . . I think this type *o* farming, if you want to call it farming, will continue to rise because the person now who owns that small tract of land *wi* a house and a *wee* bit shed is a person whose got a job elsewhere, and his wife's interested in that type *o* thing. So that's going to carry on. . . . Folk want to move out of the city; there's just no doubt about that. And quite frankly, I can see how they'd want to. . . . I think that they're honest folk that are doing it: They're doing it for the love of it. Once they get past this idea that they can make money at it—I mean, you've got to get that out of your head. You'd never make a living *aff* [off] it. (Senior auctioneer)

This chapter considers the experiences, attitudes, and behavior of people who espouse a less commercialized attitude toward farm animals. In this case, I focus on hobby farmers who typically own, breed, and sometimes show fairly small numbers of rare breeds of livestock. Although some hobbyists entertain ambitions to function at the edge of commercial livestock production, many of my contacts opted for relatively low-profit or nonprofit pursuits. Moreover, many of them enthusiastically claim to practice more traditional, natural, and welfare-friendly methods of farm-animal production than their farming counterparts in the commercial sector. The current appeal of going "back to basics," particularly in industrialized societies, has been fueled significantly by increasing public concern about the quality, methods, and safety of mainstream food production to the extent that some hobbyists slaughter and eat their own animals. I start this discussion by exploring what hobby farming means and then indicate how hobbyists see conventional farmers and how agricultural workers perceive them.[1]

Hobby farming has a long and relatively privileged rural heritage that can be traced back to ancient Chinese and Egyptian estates. Since the 1940s, however, it has no longer been the preserve of elite groups: "Hobby farmers have emerged as a significant land-owning group in many industrialized countries, including Britain and especially North America" (Layton 1978, 242).[2] For example, during the 1970s and early 1980s, many city residents in the United States relocated to rural locales, and there was considerable growth in the number of unprofitable farms with fewer than 50 acres of land.[3] Most were basically country dwellings that enabled their residents to practice hobby or recreational farming (Daniels 1986; John Aitchison and P. Aubrey, cited in Holloway 2000a, 2). Since the mid-1980s, the number of farms and ranches in the United States has contracted by 20 percent, to 2.1 million. According to statistics produced by the National Agricultural Statistics Service, hobby farms and "nouveau-farmers" account for 58 percent of all farms (Christie 2005, 1). Similarly, in 2003, "lifestyle" buyers were thought to have purchased 41 percent of farms in the United Kingdom (Mather et al. 2006, 449).[4] Following a downturn in the late 1990s, there has been a growing trend since 2004 to procure farmland for residential and non-residential use. According to the Royal Institution of Chartered Surveyors of Great Britain, this demand is partly fueled by non-farming "lifestyle" buyers, who account for 38 percent of all purchases; in some areas, this can rise to 52 percent (Commission for Rural Communities 2007, 108).[5] More recently, 80 percent of all small farms that rural estate agents sell are snapped up by newcomers to farming (France 2008, 20). Unlike commercial farmers, most hobby farmers are novices and they are part time; their livelihoods rarely depend on the productivity of their farms. The emergence of hobby farms in America stimulated concerns that, if left unchecked and under-regulated, they might "threaten the future viability of commercial farm operations by raising land prices, fragmenting land holdings, and thus hindering the expansion of commercial farms" (Daniels 1986, 31).

However, such anxieties assume that hobby farmers are purely leisure-oriented and have no ambition to expand or turn their hobby pursuits into commercially viable enterprises. This is not necessarily the case. An American farm typology drawn up by the Economic Research Service states that "residential/lifestyle farms" refers to "small farms whose operators report they have a major occupation other than farming. Some operators in this group may view their farms strictly as a hobby that provides a farm lifestyle. For others, the farm provides a residence and may supplement their off-farm income. Some may hope to eventually farm full-time" (Hoppe 2001, 4). Clearly, there are a variety of motives for small-scale farming. Similarly, Canadian research has identified and characterized two main hobby-farming types: those who are commercially motivated and regard their hobby farming as a stepping stone toward farming on a commercial footing, and those who perceive hobby farming as a leisure interest and have no commercial motives whatsoever (Layton 1978, 242). Having said all of this, one of my contacts acknowledged that her rare livestock were "mainly a hobby, but [they have] to pay [their] way; it's a small business at the same time." Despite having the vision and ambition to expand, she was "restricted by money and time." Instead of the rather static dichotomy of motives outlined above, we could envisage a more fluid and dynamic continuum. For example, changing socioeconomic circumstances, at the individual and structural level, as well as public attitudes toward large-scale farming, might mediate the extent to which hobbyists can realistically make the decision to shift from hobby farming to more financially oriented farming.

Some part-time farmers are mainstream farmers on the road to retirement: "Part-time farming has frequently been described as a transitional stage for families on the way into, or more likely out of, full-time commercial agriculture" (Gasson 1988, 120).[6] Plainly, there are different reasons that people farm on a part-time basis (Gasson 1990). Unlike ordinary farmers who might engage in part-time farming as a strategy to exit or convert from mainstream agriculture (Gasson 1988), hobby farmers enter part-time farming as a lifestyle choice that may or may not have the potential to evolve into a commercially viable business. Some have argued that hobbyists are a distinctive subgroup of part-time farmers (Daniels 1986). As new entrants into farming, they are typically middle-aged and formerly urban, and they have pursued other careers or occupations. They have also purchased their smallholdings as opposed to inheriting or renting because the "economic bonds and family ties with the land or . . . particular holding which characterise long-established farming families are absent" (Gasson 1988, 135). Given all this, I now consider the status and credibility of hobby animal farming as perceived by those involved in conventional agriculture and how hobby farmers themselves make sense of their activities in relation to mainstream livestock production.

"The Good Life"

A growing number of people, recently dubbed the "greenshifters," are leaving the bright city lights to live in the country (France 2008).[7] Although this migrational trend of urban to rural is not new, it has been "described as a third agricultural revolution. A new generation of people living—as much as they can—off the land: swapping their inner-city gardens for hardscrabble smallholdings, the daily commute for early-morning goat milking, their domestic cats for pedigree pigs, their Blackberries for home-made compost"

(France 2008, 20). According to one manager of a rare-breed farm who used to farm commercially, hobby farmers are "people who don't really need to make a living off of it. You know, it's sort of part-time farmers, smallholders, people who've got other jobs and they're doing it . . . as [a] sort of release more for themselves and the pressures of modern life rather than doing it to make a living." Similarly, a married female hobbyist said that a "hobby farmer" was "somebody that makes not a bloody bit of money out of the whole thing. . . . [And] you anthropomorphize your animals." She typified her vision of hobby farming by recounting a couple of anecdotes:

> I've got two sheep out there. They're both wedders.[8] One was born two years ago, and it was born from incest—accidentally, obviously. One of the young lambs had gone mad . . . and impregnated all the ewes that year. We were lambing in January, and this thing was born, and [my husband] saw it lying dead, [as] he thought. He actually started to dig a hole for it and saw a little foot twitch. He brought it to me, [and] for all the morning I kept it in warm water in the sink. It was tiny . . . and it survived, and it cost me a lot of money at the vet's because it wouldn't suck. We had it in the house for over a week, in the bedroom. He survived, and he's out there with . . . a big cow bell around his neck. He's called Lazerth. [He has] the bell around his neck . . . because the next year we had a pet lamb; its mother wouldn't feed it. It wasn't quite right, and after a week we realized that it was blind. So we've got Lazerth leading blind Billy, 'cause Lazerth got a bell around its neck. That's hobby farming!

She then told me what she meant by the terms "livestock" and "pet" but proceeded to explain what she thought livestock should mean: "It should mean that a pet is something that has no financial value whatsoever, and your livestock should make money. You should get some money back from livestock. But both of us have . . . crossed that barrier, and the livestock are pets, as well." This deviated from what she regarded as the "traditional view" of having livestock, which she equated with farmers who make money from their animals: "We're not farmers. . . . We can be a bit of a joke, because we've got rare sheep and they've got horns, and that's bad. I've got a Highland cow—who on earth has a pet Highland cow that you never put in calf?" So some hobbyists might see themselves as playing at farming. Due to relatively privileged personal and financial circumstances, their incomes are minimally linked to the productivity of their animals (if they are linked to it at all). In other words, they can afford to treat their livestock as pets (see Holloway 2001).

In a world where herd size matters, the fact that most of my contacts kept fairly small numbers of rare livestock did nothing to enhance their already fragile credibility with their farming neighbors. A female hobby farmer illustrated this: "If I put something that's spots and horns or something that's a fuzzy little teddy bear through market, they would just, phew! Because it's not a Suffolk or a Blackface . . . they don't want to know; you wouldn't get the price for it. But the fact that it's going deadweight, once it's eaten, nobody cares what it looks like—or did look like." As already noted, commercial sheep can be penalized, too, especially hill breeds. This indicates a species hierarchy of live sheep breeds. Lowland commercial breeds are financially and aesthetically valued over commercial hill-bred sheep, and rare breeds are designated the least commercially viable grouping, especially in the livestock-auction setting. However, as I discuss shortly, breed hierarchies are not fixed in stone.

It would seem that hobby farming per se is not taken seriously by the commercial sector. For example, a senior auctioneer who was sympathetic to why people might be drawn to such ventures also doubted its validity by talking about "this type o farming, if you want to call it farming." Given that, in agricultural circles, full-time farming is regarded as the normative benchmark and is superior to part-time farming, it is hardly surprising that the juxtaposition of the terms "hobby" and "farming" appears, at best, to be an oxymoron and, at worst, anathema.[9] Even some hobby farmers want to distance themselves from the term. David Walbert, a writer and historian who lives and farms in North Carolina, suggests that hobby farming "implies dilettantism, the tendency to dabble in this and that without effort, knowledge, or gain. To call small-scale agriculture 'hobby farming' makes it nearly impossible to take that activity seriously" (Walbert 1997–2004, 2).[10] Instead, Walbert prefers "amateur farmers." Although he is aware that "amateur" conjures up negative associations of "sloppy work, work based on too little knowledge or understanding," it also refers to someone who really loves his or her work and will make every effort to do it as proficiently as possible (Walbert 1997–2004, 7–8). Interestingly, many hobbyists also dissociate themselves from the term "farmer." For example, a married rare breeder, who reversed roles with his wife to stay at home with their children, explained that he did not regard himself as a farmer because he doesn't "make any money out of it." Despite being described as a farmer in the local press, he insisted that he saw himself as "someone fiddling around . . . but dealing with all the same issues that people who are farming properly have to deal with." He realized he and his wife had "a privileged kind of start" because they had "no mortgage" and knew they "had the ability to earn money" outside of their animals. Despite friends in London characterizing them as "good lifers," his wife commented that the so-called good life "is bloody hard work!" The wife of a middle-aged professional, who is trying to be as self-sufficient as possible, explained, "It's like having a baby—you have no idea how hard it's going to be until you do it" (France 2008, 26). A young couple shared the sentiment that hobby farming was "damned hard work" and similarly did not regard themselves as farmers: "We're lucky enough to have a bit of land to run enough animals to provide us, family and friends with organic free-range produce. We enjoy doing it, but I [the husband] wouldn't class it as farming." Finally, a middle-aged hobby couple, who had given up their professional occupations and moved from England to relocate in Scotland, acknowledged that "compared to real farmers around us, we play at farming really." Even though some of their livestock were "paying [their] way plus a bit," they were aware that, overall, they did not depend on their animals to survive financially. They also perceived themselves as "on the fringes of farming: We're within a farming community, but not part of it." The husband acknowledged that he would prefer to be less peripheral and explained, "There are elements . . . of that farming community that we depend upon."

This dependence on experienced farmers can be double-edged. For example, farmers are usually willing to extend a helping hand to support their inexperienced hobby neighbors with bailing hay, shearing sheep, fencing, or giving general advice about husbandry. One of my rare-breed contacts said that her need for help from a neighboring farm actually enhanced the farmers' "respect" toward her and her husband. By admitting her inadequacy, she was implicitly acknowledging and deferring to the greater level of knowledge and experience of the conventional farmer, which the farmer clearly appreciated. On other occasions, farmers tried to persuade hobbyists to be more conventional and not squander their time on non-commercial breeds. A male hobbyist warmly recounted that a retired farmer had told him he was "wasting [his] time keeping these sort of sheep" and that he should "get rid of these goats." Instead, he ought to get "a few ordinary sheep . . . that everybody else keeps, then you can take them to market, and there won't be any problem." Finally, a hobby couple who had some misgivings about the activities of farming neighbors were unwilling to "rock the boat." They were "surrounded by much bigger farms with much more clout and much more machinery," they explained, and although they were concerned by what they had seen, the husband asserted that he had no desire to stir up "animosity" because he needed "to get on with [his] neighbors."

These examples indicate how hobby farmers, many of whom are not only incomers settling in fairly well-established agricultural communities but also new entrants to farming, might be considered by ordinary farmers as city outsiders who are naive and ignorant of farming ways.[11] Unsurprisingly, many "good-lifers" function, and perceive themselves, as on the periphery of "real" farming. This is materially and symbolically reinforced by the "alternative" livestock breeds, farming worldviews, and husbandry practices they champion. Moreover, while producing and showing animals primarily for pleasure undoubtedly offers its enthusiasts an alternative kind of life, it could be the case that hobbyists are simply more intent on "owning land than farming it" (Gasson 1988, 114). For this reason, non-productive factors such as scenic beauty might be more influential in terms of where they decide to stay (Layton 1980, 221). This contrasts strongly with commercial livestock farming, in which livestock are the "tools of the trade" through which producers secure their livelihoods. Having said all this, farmers are amenable to assisting hobbyists and will readily impart agricultural pearls of wisdom. The physical act of doing so provides an important, albeit tenuous, connection between their two worlds. Ian Dalzell, who oversees countryside membership services for the National Farmers Union, suggests that the increasing trend of people seeking the rural life provides an opportunity to "reverse the criticisms of 'weekenders' who are traditionally seen as contributing little to the economy. The criticism used to be that people moved to the countryside simply because they wanted a chocolate-box existence . . . and that places ended up as dormitories from Monday to Friday. . . . Our smallholders are working and buying locally so they take a more active part in the local economy and the local community" (Dalzell 2000, 1–2). Moreover, in 2000, two third-generation Derbyshire dairy farmers, Robert Jeffery and David Morris, formalized and commercialized this exchange by becoming consultants to those "who are new to agriculture, conservation and rural land ownership" (NewLandOwner 2008). They "realised they had something more marketable than milk—their experience" and set up NewLandOwner (France 2008, 20), a smallholder's consultancy service that offers advice on a wide range of farming and land-related matters, as well as a number of training courses. For example, NewLandOwner runs a weekend-long "Getting Started" course for smallholders in which participants get as much hands-on experience as possible and practical guidance from experienced local farmers and former agricultural lecturers on issues related to smallholding. They also run a one-day butchery course that shows smallholders how to turn livestock carcasses into edible, and possibly saleable, cuts of meat. As Jeffery and Morris state on their Web site, "Whatever your aims . . . we have the expertise to ease you into your new lifestyle."

Rare Breeds of Livestock: Paucity and Purity

At the start of the twentieth century, there were about 230 native breeds of cattle throughout Western Europe. "By 1988, only about 30 of these breeds were secure; 70 had become extinct; and 53 were in an endangered state; while the remainder were minority breeds, not immediately endangered but far from secure" (Dowling et al. 1994, 14). The situation in America may be rather different because "most American domestic livestock is descended from imported ancestors and there are few genuinely native breeds" (Alderson 1994, 25). Even so, in the late 1920s the U.S. government took steps to safeguard the Texas Longhorn by providing the financial backing to establish herds in Oklahoma and Nebraska. While the breed originates from, and can be traced back to, Spanish cattle introduced into America, the Texas Longhorn is so closely associated with the "romantic history of the cattle trails and pioneering of the Wild West that it is accepted as an American Breed" (Alderson 1994, 26). Specialists at the International Livestock Research Institute and the Food and Agriculture Organization have identified the "globalization of livestock markets" as one of the most significant factors contributing to the declining diversity of livestock breeds: "Most of the world's rapidly growing demand for livestock products is being met by intensive production systems based on a few species and breeds of high-input, high-output animals. Intensive production systems often bring with them erosion of local animal genetic resources" (International Livestock Research Institute 2007, 32). Countries doggedly pursing the intensive-production route experience the most prevalent demise in their breed categories because farm animals considered incapable of meeting the commercial demands of production tend to fall out of favor (Alderson 1994).

The British picture is no exception. "Livestock farming has always been a fashion industry. . . . For instance, in the '60s, the Beef Shorthorn breed of cattle was very popular for the 'baby beef' market, as it was then. Today it is a rare breed" (Lutwyche 1998b, i). The Rare Breeds Survival Trust (RBST) claims that during the twentieth century no fewer than twenty-six indigenous farm-animal breeds became extinct in Britain, but none have done so since 1973, which coincides with the inception of the trust (Evans and Yarwood 2000).[12] The RBST's raison d'être is to actively monitor and promote the genetic conservation of rare breeds of domestic farm animals. The aims and vision of the RBST stimulated like-minded groups to set up sister organizations throughout the world, such as the American Minor Breeds Conservancy (1977), the Joywind Farm Conservancy in Canada (1986), the Rare Breeds Conservancy in New Zealand (1989), and the Australian Rare Breeds Reserve (1990). Rare Breeds International was set up in 1991 to oversee the activities of these and other groups at a local and global level (Alderson 1994, 20–21).

The criteria adopted by the RBST to describe degrees of rarity were originally drawn up by scientists at Reading University (Evans and Yarwood 2000).[13] There are currently six categories of priority: critical, endangered, vulnerable, at risk, minority, and other native breeds (Rare Breeds Survival Trust 2008c).[14] Genetic assessment of purity lies at the heart of legitimate inclusion or exclusion, along with the total numerical population of a particular breed: "When a breed population falls to about 1,000 animals, it is considered rare and endangered" (RBST n.d.). As the numbers of a rare breed increases, the breed moves from the trust's priority list to its minority list. This ensures that the trust will continue to assess the breed's rarity status. However, the trust could become a victim of its own success, because as rare breeds become increasingly reestablished, it throws into the question the nature of continued input, if any, from the RBST. It has been suggested that the "status as a 'minority breed' demands RBST involvement with the breed through monitoring. This strategy to retain influence was replicated in . . . 1996 when a new category of 'rareness' was introduced, the 'native breed'" (Evans and Yarwood 2000, 239).[15]

The RBST's rationale for creating the new classification of "native breeds" (nucleus stock) related to concerns that "some native breeds are suffering from introgression to the extent that modern stock bears little resemblance to the original breed" (Rare Breeds Survival Trust 1996, 55). For instance, the genetic purity of British-pedigree Hereford cattle was causing unease because the trust believed this particular bloodline was being diluted by the introduction of genes derived from American Hereford cattle. It contends that the American breed is bigger in stature than British Herefords because the American Hereford is cross-bred with Simmental cattle (Evans and Yarwood 2000, 239). To assess the extent of introgression, the RBST analyzed about 2,000 blood tests taken from British Hereford cattle and concluded that only 350 of the cattle could be regarded as "pure British Herefords not showing signs of alien influence" (Rare Breeds Survival Trust 1996, 55). The Hereford Cattle Society rejected the trust's indictment (Evans and Yarwood 2000, 239), even though a high-profile Hereford breeder allegedly conceded that crossbreeding does occur (Rare Breeds Survival Trust 1996, 56). Regardless of the dispute surrounding its genetic assessment, the RBST decided to distinguish the "traditional" British Herefords by giving them a "special status" in the Hereford Herd Book (Rare Breeds Survival Trust 1996, 55–56). Clearly, assertions of rarity, purity, and nativeness have become integral to claims of a breed's authenticity. But how might non-genetic factors such as animal husbandry and productive practices also influence perceptions of how strictly animals epitomize ideal characteristics of a breed or species?

The following example taps into debates within the North American bison industry (Lulka 2006, 2008). This fledgling industry is currently grappling with the pros and cons of moving away from hobby-bison ranching to establish a commercialized approach to bison-meat production. This has triggered heated disputes over methods of producing, feeding, and slaughtering bison that have raised fundamental questions and expressions of concern

as to how these practices might alter the "physical and behavioural characteristics of the species . . . in the process of establishing a market for bison
products" (Lulka 2008, 52). One of the main reasons hobbyists are particularly attracted to conserving bison is the historical-cultural resonance of wildness these animals hold in the American imagination. Before 1875, colossal
herds of wild bison freely roamed and physically dominated the prairie landscapes; however, their fate was sealed as soon as a market emerged for their
skins and hunters cashed in on this growth area. The arrival of the railways
during the 1860s "drove a knife into the heart of buffalo country," and by the
1870s, improved tanning procedures meant their skins could be turned into
a pliable and most desirable source of leather (Cronon 1991, 215–217). The
wholesale slaughter of bison during the nineteenth century decimated "the
North American population of 30 million bison . . . to approximately 1000
animals" (William Hornaday, quoted in Lulka 2008, 31). From the standpoint
of simply "reintroducing" bison or increasing their numbers, most would
agree that this goal is being realized; recent estimates suggest there are now
more than 200,000 animals (Lulka 2006, 175). This resurgence in numbers
coincides with an upsurge in interest in bison ranching during the 1990s
and a rocketing demand for bison.[16] The industry responded by setting up
a breeders' market. This market supplied the burgeoning numbers of novice
hobby enthusiasts who wanted to keep these emblematic wild American animals. It also enabled established producers to capitalize on the soaring prices
paid for bison by moving into breeding to sell their calves. Eventually, supply
outstripped demand, and the price of bison slumped (Lulka 2008, 38). When
the bottom fell out of the bison market, the industry was forced to reflect on
how best to proceed. One perspective was voiced by a spokesperson of the
National Bison Association (NBA): "On the one hand, the NBA and bison
producers must continue to promote the romance surrounding the American
bison. Mystique marketing is a strong asset for our business. On the other
hand, we need to de-mystify bison so that consumers quit associating bison
with the meat that is only eaten on special occasions in exclusive restaurants"
(Carter, quoted in Lulka 2008, 40).

Calls by some groups within the industry to conventionalize bison-
meat production tend to prioritize financial considerations at the expense
of cultural-environmental concerns or the animal itself. This is out of step
with the principles associated with "bison restoration." Although similar to
"reintroduction," which aims to increase the bison population, restoration
also "stresses the quality of wildness and the agency of non-human animals"
(Lulka 2008, 33). As mentioned, wild bison are closely associated with the
landscapes they liberally roamed and grazed. Commercially farmed bison are
more likely to eat grain than grass and to be enclosed on private land, or even

placed in feedlots, than to rove open spaces. If bison are reclassified as an "amenable species," like domesticated cattle and sheep, this could have implications for how bison are slaughtered and inspected in the future. As some of David Lulka's interviewees note, such changes "are transforming the nature of bison rather than restoring its qualities. . . . [T]he NBA and the bison 'industry' through . . . practices and procedures . . . are in essence changing bison into just domesticated furry cattle" (Lulka 2008, 44, 51). This indicates that increasing the actual numbers of endangered species is only part of the conservation process. If repopulation is not accompanied by more ecological and animal-orientated aims of restoration, the authenticity of the beast will come into question. What the argument about preserving the bison shows is that an animal is always more than its genes. In this case, a character is assumed from the way the animal used to behave in a specific environment. Bison roamed the prairie, so if it stands in a feeding stall, it is not a "real" bison.

"A Chance to Survive": Rare Breeds and "Post-productivist" Agriculture

In 1998, the RBST celebrated its twenty-fifth anniversary. In recognition of this, HRH The Prince of Wales, Patron of the Trust, wrote the following message in the trust's journal, the *Ark*:

> In 1973 "conservation" was a word which was hardly ever heard. In farming circles, rare breeds were too often considered to be near-dinosaurs which had served their purpose and which could now be abandoned to their fate while more "modern" types took their place. . . . Rare breeds are now recognised by farmers far and wide to have a worth and a function, and a value for purposes, perhaps not yet thought of, which makes their conservation necessary as well as desirable.[17] (HRH The Prince of Wales 1998, 93)

Clearly, these decades signal a volte-face of agricultural opinion and coincide with a period of considerable restructuring within the U.K. agricultural sector. The key aim of agricultural policy during and after World War II was to encourage farmers to produce as much food as possible. Because Britain had experienced serious food shortages and rationing during the war, when food imports were severely curtailed, all stops were pulled out to make the country self-sufficient. This bolstered "the productionist paradigm" and legitimated the industrialization and intensification of farming and food production in Britain. This model is epitomized by the shift away, although not absolute, from "local, small-scale production to concentrated production and mass distribution of foodstuffs" (Lang and Heasman 2004, 18–20). Furthermore, during the war the British government provided a range of financial incentives and subsidies to encourage farmers to make necessary changes and improvements to their land to increase food production. Although many farmers went to extraordinary lengths to realize this goal, the intervention of assured prices for agricultural commodities by and large compensated farmers for their efforts (Hodge 1990). By the end of hostilities in 1945, farmers, who normally were fairly conservative in their politics, were positively inclined toward government intervention in the market, even if it meant state direction of their work. The passing of the Agricultural Act in 1947 did much to safeguard continued support for the British farming sector for many years to come.

Clearly, farming interests, including influential bodies such as the National Farmers Union, had significantly contributed to and shaped domestic food and agricultural policies. However, when the United Kingdom joined the European Economic Community (EEC) in 1973, it had to adopt the Common Agricultural Policy (CAP).[18] The EEC, or Common Market, was ratified in 1957 when the six founding member states—Belgium, France, Germany, Italy, Luxembourg, and the Netherlands—signed the Treaty of Rome. One of the overriding aims of the treaty was to facilitate unfettered trade between these countries. The original objectives of CAP were in keeping with postwar agricultural policies drawn up by the United Kingdom and other industrialized Western countries, including the United States: to boost production, stabilize markets, provide moderately priced food to consumers, and, most important, increase farmers' incomes "to ensure a fair standard of living for the agricultural community" (Howarth 2000, 4). However, subsidies intended to support farm incomes encouraged the production of unwanted food surpluses. By the mid-1980s, surpluses had peaked, and such policies had triggered a crisis within farming:

> Farmers were offered subsidies to specialise in certain types of food production, use more machinery, buy more products to feed livestock and to use more artificial fertilisers. Although they improved food output, these intensive methods caused environmental damage and led to a "farm crisis" caused by overproduction of foodstuffs, the immense budgetary costs of agrarian support and falling farm incomes. (Yarwood and Evans 1999, 80)

The quest for efficiency put many farmers "on a technological treadmill" (Evans and Ilbery 1993, 945–946). For example, increasing reliance on capital loans to finance their businesses accentuated the need to intensify farming practices, often to the detriment of the environment. Some responded by expanding and investing further in their agricultural firms, while others, principally small-scale family farms, did their utmost simply to survive. For many, this meant diversifying into a range of on- and off-farm business ventures, such as bed and breakfast accommodation, farm shops, and rare breeds of livestock (see, e.g., Ilbery 1988, 1991; Evans and Ilbery 1993). Throughout the industrialized world, the small family farm was threatened. For example, in the 1930s there were about 7 million farms in the United States; by the mid-1990s, there were fewer than 1.8 million. During the 1960s, France had 3 million farms; this fell to about 700,000 in the 1990s. Between the 1950s and the 1990s in the United Kingdom, the number of farms contracted by nearly 50 percent, from 450,000 to approximately 250,000 (Lang and Heasman 2004, 149).

By the 1990s, it was no longer desirable or, indeed, sustainable for farmers to diligently churn out copious and superfluous amounts of food. Subsidies had been reduced, and the link with output had been broken. Issues that previously had been relegated to the periphery of agricultural policy, such as the sustainability of the environment and of rural communities, increasingly took center stage. These policy revisions promote flexible and diverse approaches to farming while encouraging environmentally sensitive projects. This could be described as a dilution of productivist aims and objectives and the emergence of "post-productivist," or even "non-productivist," type of agriculture.[19] But the farming sphere is rarely as neat as such classificatory frameworks. As Lulka (2006, 182) notes when writing about the bison industry in America, "Is it the animal, the mode of production, the quality of food, the industry in relation to other industries, the scale of operations, or the larger native context (to name a few variables) that determines the appropriate and final classification of an industry?"[20] The point is well taken, but terms such as "productivist" and "post-productivist" still signify vital differences, and I continue to use them throughout this discussion.

Recent changes in agriculture have opened up new opportunities for rare breeds of livestock. Although unwanted when maximizing output was the main concern, "They have found four new roles within the post-productivist era" (Yarwood and Evans 1999, 81).[21] These include niche markets for rare meat, family farm parks, symbolizing local and national rural heritage, and improving the environment. In addition, rare breeds often require little intervention, which makes them attractive to time-pressed hobby farmers. I now illustrate each of these aspects in more detail.

As current concerns about obesity demonstrate, modern economies no longer struggle to produce sufficient calories to maintain health. Food in general, and meat in particular, has become so cheap relative to average earnings that significant proportions of our populations can afford to be selective.[22] They can expect their food to be satisfying in more than just the physical sense—hence, the creation of niche markets that promote rare breeds as superior to the commercial varieties for taste, animal welfare, the environment, and heritage preservation. Just as the bison is promoted as Americana, so the RBST says, "If you want to see British Breeds in British fields you've got to eat them!"[23] Although the trust was keen to set up a "farmer to butcher distribution chain to create a market for rare breed meat products," it was uncertain how members of the general public would react to this specialist meat-marketing scheme (Rare Breeds Survival Trust 2008b, 1):

> One of the hardest policies to implement was the Traditional Breeds Meat Marketing Scheme. On the one hand, our meat producing rare breeds desperately wanted a commercial outlet to make keeping them viable. On the other hand, it was widely thought that the public would throw up their hands in horror at the thought of eating anything labelled "rare." . . . Not a single customer turned away from the prospect of eating rare breeds.[24] (Lutwyche 1998a, ix)

Very few people can afford to "keep rare breeds just for their own sake—for conservation," said a female rare breeder, which reinforces the RBST's strategy of commercializing rare breeds to make them more viable. As noted earlier, some hobbyists expect their animals to pay their way, which clearly indicates that a productive ethic can be present in post-productivist farms (Holloway 2002, 2060). As Robert Goodin suggests, "Post-productivists are not opposed, or even indifferent, to economic output. . . . [T]hey have simply 'gotten over' being utterly fixated on it, as productivists have been" (quoted in Mather et al. 2006, 451). The RBST currently authorizes forty-five independent butchers to sell rare-breed meat (Rare Breeds Survival Trust 2008b).[25] This is consistent with the trust's philosophy that rare livestock and their products should find an alternative market to the mainstream markets, especially supermarkets. This is an important point, given that the four big U.K. supermarkets have three-quarters of the total retail grocery market. Proceeds derived from the retail meat sector are in the region of £7 billion per annum. Clearly, "Meat is big business" (Meat and Livestock Commission 2008, 6). Retailers have carved up this sector into a number of price tiers whereby products might be promoted as, to name some of the main options available, "value," "standard," "premium," or "organic." Predictably, most supermarket consumers bought meat products from the standard price tier, as opposed to the premium tier. Hence, this accounted for 65 percent of beef sales, while only 4 percent of customers bought premium beef products (Meat and Livestock Commission

2008, 15). As the balance of power in the food-supply chain has noticeably shifted away from producers toward consumers and retailers, what consumers want, consumers generally get (Lang and Heasman 2004). To cater to the bulk of their customer base, supermarkets are looking primarily, though not exclusively, to sell standardized animal products that have been intensively produced for a mass market and that can be retailed as cheaply as possible. Most rare breeds are extensively produced and are less likely to meet the standardized carcass-grade specifications for which supermarket buyers are looking. Given this, some note that "supermarkets are looking for a consistent product which is lean and easy to handle. They cannot countenance unusually small breeds or their opposites, or breeds which carry a greater fat covering, even though that meat will have much more flavour and succulence" (Lutwyche 1998b, ii).

Rare-breed meat is presented as a "high-quality" product that is traditionally produced and skillfully prepared by accredited specialist butchers. "By bringing together their traditional skills to the proper preparation and hanging of the meat, all the tenderness and juicy succulence will be brought out when the meat is lovingly cooked—something you will no doubt appreciate when you sit down to dinner" (Rare Breeds Survival Trust 2008b).[26] The trust draws on nostalgic notions of how flavorful farm-animal produce used to be by suggesting that, if you eat rare breeds, you will get pleasure from eating "meat with real old-fashioned qualities of taste and succulence" (Rare Breeds Survival Trust 2008b; see also Pollan 2006). Customers can also be assured that such meat is fully traceable; that it has been produced locally using "a more natural way of farming," and that they are doing their bit to conserve rare breeds. Such narratives appear to resonate with discerning consumers. Indeed, a rare breeder actively involved with her local RBST branch maintained that some consumers were able to appreciate the tastes and textures associated with different breeds of rare meat. Buying accredited rare meat enables such customers to demonstrate not only their environmental, conservation, and animal-welfare-ethics credentials but also their cultural distinctiveness and finely attuned taste buds. In other words, rare-breed meat has become a form of culinary capital and a form of social distinction.

More than two-thirds of my hobby contacts preferred meat produced from their rare livestock, or fellow hobbyists' stock, to supermarket meat. A rare-sheep breeder who supplied a local restaurateur claimed, "Anybody who comes here [her home] and eats the lamb I produce say it's the best lamb they've ever tasted." She also explained, "There's a terrific scorn among the commercial people" toward her animals. "I point out this is a non-commercial animal which has a far better flavor than any Suffolk and Texel crosses," she said, but "nobody would dream of producing [Ryeland sheep] commercially because they're not fashionable." In fact, "If [commercial farmers] got a flock of Ryelands, their street cred would go down enormously." Other hobby farmers produce and eat their own animals because they distrust the quality and safety of mainstream food production. A young mother attempting to be self-sufficient said, "There seems to be a different scare [each week] about the way food is produced in large quantities. And every time you hear something else, you think, 'That's something else you have to strike off your list, that you can't buy.'" Her main concern, she said, was to safeguard her children's health by keeping their food as "pure and natural as possible." She said she would "rather eat something natural and unprocessed" than a processed meal that "has so many additives and things in it that you've no idea of what [it will] eventually . . . do to your system." By producing their own food, she felt, the couple knew exactly what they were feeding their children. This narrative not only assumes that healthy animals = healthy food = healthy people; it also shows that "major threats to animal health, and by association human health, are held to come from the practices of modern, intensive husbandry" (Buller and Morris 2003, 226–227).

The desire to adopt more traditional methods of production has motivated a new generation of hobbyists to roll up their sleeves and acquire the animal-husbandry skills required to keep and produce rare breeds of livestock. For example, John Seymour's *The Complete Book of Self-Sufficiency* (1996), which explains and promotes the principles of food self-sufficiency, has sold more than 650,000 copies in thirteen countries since it was first published in 1976. Many hobby farmers say that their approaches are more ethically informed and respectful of the farm animals than conventional production systems. Mary Marshall, a representative of the Smallholders' Forum, said that hobby farmers "often spend more on proper prevention practices than do commercial farmers, and tend to have more veterinary involvement and spend more time per animal" (Elliot 2008).[27] Similarly, it was important for my hobbyists that their animals had a "natural" and "good" life. A male hobbyist explained that his animals "lead an active natural life, and they're fed entirely off my own croft." A couple asserted that their farm animals "generally have a very good life when they are here." The wife then elaborated by saying how "proud" she was of this because it was important that their animals were "looked after properly." She said this in the context of a discussion in which she noted that she became irritated when people questioned how she could eat her animals because it made her "sound cruel." However, she defended her position by suggesting that those people were hypocritical, overly sentimental, and not fully informed about "how animals are treated." She assumed the high moral ground by believing that she and her husband had done everything they could to ensure that their animals had not suffered and had enjoyed a "good life." This position has been given ample publicity in a popular British television series presented by Hugh Fearnley-Whittingstall, a celebrity chef and smallholder. In an interview he explained that he had no objections to animals' being reared and killed for meat so long as the animal's life had been "reasonably comfortable and fulfilling" (Richards 2008, 1). Although most rare breeders were sympathetic to Fearnley-Whittingstall's viewpoint, a few said they did not eat their own livestock because, as one put it, "It would be like eating [my] dog."

Even though most hobby farmers are animal-centric and often articulate welfare-friendly production approaches, varying levels of ignorance, inexperience, and lack of time can inadvertently cause harm to their animals (see Farm Animal Welfare Council 2007). The ignorance can be quite profound, especially when the hobbyist is starting out. A former nurse told me, "I bought four ewes and then realized . . . that I would need a ram. I don't know how I thought it was done." She was aware of her limitations and addressed them by reading books and by "asking people, watching, and doing it." She also drew on her nursing experience. "There's not that much difference between our ailments and their ailments," she said. "They have bones that break; we have bones that break. So I can usually go back to a medical experience and treat them accordingly." As mentioned earlier, introductory practical courses are an important resource for new smallholders to become better informed and gain more confidence in animal-related responsibilities. One of my male contacts decided to attend such a course when he acquired goats and hens. He took the course primarily because he knew he would have to "kill a hen and . . . pluck it and clean it and get it ready for the table. . . . [U]nless I do this in a group, I'm never going to be able to do it." An experienced hobby breeder said that general levels of technical knowledge were low: "There's not enough, I know I'm being big-headed here, but knowledgeable people in the rare breeds . . . a lot of them, their heart's in the right place, but they're just not too knowledgeable about handling livestock and being involved in livestock." She tried to address the skills gap in her RBST group: "In the wintertime, we have a program of meetings and things that people come along to. . . . We have vet talks and feed companies [to give] the nutrition side of things. . . . Hopefully it'll help them a bit, but basically most of them do need a *wee* bit more information." Such knowledge gaps are not confined to Scotland or to this hobbyist's backyard. The Ministry of Agriculture and Forestry in New Zealand had also received a record number of complaints about the welfare of livestock owned by hobby farmers. Ross Burnell (1999, 1), the spokesperson for the ministry's Enforcement Unit, explains:

It is common for hobby farmers to own several types of livestock and they sometimes do not fully appreciate the requirements of all the species in their care. . . . Most livestock owners mean to care for their animals well, but good intentions are not enough for good farming. Inexperienced livestock owners must not risk their animals' welfare but must seek advice from an expert if they are unsure of good farming practice.[28]

However, juggling hobby farming with other part-time or full-time work can limit the amount of time hobbyists can realistically spend checking their animals. For example, a rare sheep breeder explained that she had been "a bit more neglectful because my husband and I parted company about three years ago; since [then], I've been out at work full-time. So I can't do as much with them as I used to. So there's the odd one wandering around lame. . . . I really just haven't had time to go and do anything about it, whereas normally I would have paid attention regularly to the state of their feet. . . . So it's not ideal." Despite earnest attempts to look after farm animals properly, in practice this can be easier said than done in both hobby and commercial contexts. As noted by the National Farmers Union (2008b, 1), "There may be a small minority of cases in which farm animals are being kept by people who, for whatever reason and on whatever scale, may not be well-equipped to identify disease and other risks. . . . How to address that situation, in ways that do not add to the already huge regulatory burden on livestock farmers of all shapes and sizes, needs to be the subject of debate." Although some groups within and outside the commercial sector might dispute the extent to which the welfare and rights of farm animals have been, and realistically can be, addressed in productive settings, the fact certainly remains that livestock continue to be bred, reared, and slaughtered for human consumption. Sensitivity costs. But too much insensitivity costs too and is illegal. Market conditions and profit margins fluctuate, as does the economic value of livestock. The ethical and productive ingredients underpinning livestock farming make up an unstable and potentially paradoxical cocktail. At the end of the day, most producers will have to grapple with the challenge of how best to reconcile caring for their animals while balancing their books (Wilkie 2005).

Welfare is one element of animal farming. Another is physical distinctiveness. In the case of rare cattle breeds, the visual appearance of these animals can greatly influence their chances of becoming extinct: "If the Irish Dun and Suffolk Dun had been a more spectacular colour they might have escaped extinction" (Alderson 1994, 71). Thus, it would seem that conservation decisions can be contingent on arbitrary, non-genetic factors that cannot always be uncoupled from the aesthetic characteristics of the breed in question. Indeed, many endangered breeds are visual spectacles. Gloucester Old Spots pigs, Norfolk Horn sheep, and Vaynol white cattle with black markings on their noses, ears, and feet are ideal candidates for family farm parks because their striking markings catch the public's attention. The manager of a council-run rare-breed farm said he selected animals that are visually different from "average commercial animals" because this helps visitors identify and differentiate between the various breeds. Although he thought his council bosses were not necessarily interested in rare breeds, they had supported the project because "it fitted in with their conservation image." A number of heritage and conservation organizations, such as the National Trust in Britain and the Joywind Farm Rare Breeds Conservancy in Canada, have also set up farm parks (Alderson 1994, 70), and the RBST currently approves sixteen such parks throughout the United Kingdom (Rare Breeds Survival Trust 2008a). Ordinary farmers have also been attracted to creating smaller-scale breed parks as an opportunity to diversify into alternative farm-related projects as a way to supplement declining farm incomes. Such initiatives received financial support between 1988 and 1993 from the U.K. government through the Farm Diversification Grant Scheme. A "small but significant" number of farmers submitted proposals to the scheme, which indicates that "there was a realisation, albeit for commercial reasons, that certain breeds were worth preserving. 'Unusual' animals have a novelty value that appeals to visitors but which can shatter local landscape coherences" (Evans and Yarwood 1995, 144).

The reference to "local landscape coherences" leads to the third role of rare breeds: Such animals have become symbolic signifiers of rural locales and identities. As Richard Yarwood and Nick Evans (1998, 155) point out, "The shift in agriculture has caused rare breeds to be revalued, not for their productivity, but for their role in the sanitized reconstruction of past rural life. Gloucester Old Spots pigs and other rare breeds are finding themselves in new types of rural spaces that are bound up with consumption rather than production."[29] In other words, rare farm livestock have become animated icons of our rural heritage. As HRH The Prince of Wales has put it, "Our pedigree farm livestock is just as much a part of Britain's heritage as is her castles, her art collections or her historic churches!" (quoted in Rare Breeds Survival Trust 1996, 56). Consider the findings of a 1996 survey of the distribution of three native Irish breeds of cattle (Yarwood et al. 1997).[30] Perhaps unsurprisingly, two of the three breeds, the Irish Moiled and the Kerry, were most prevalent in their geographical area of origin, the north and south of Ireland, respectively.[31] In contrast, Dexter cattle were common on the British mainland, not in southwestern Ireland, from which the breed stems. Yarwood and his colleages (1997, 21) suggest that "a third of all Dexter and Kerry cattle were kept as visitor attractions, as were half of all Irish Moiled cattle. Consequently, half of the respondents felt that rare breeds were important for national heritage or, at least, a commoditized version of it. In these cases, animals were kept because 'they were Irish.'"[32] In contrast, Dexter cattle were favored for their productive characteristics, not their ethnicity. Some Irish breeds have been given a boost as they are closely associated with, and perceived to be representative of, a rural location's identity and its heritage.[33] Clearly, rare breeds are appealing partly because they can be a visible sign of regional or ethnic distinction and partly because they are distinctive. The long, shaggy coat and great horns of a Highland cow add "Highlandness" to the rugged landscape of Sutherland or Skye. But they also add "cowness" to cows. The renaissance in native rare breeds may be a cultural reaction to the proliferation of continental breeds that increasingly pervade local, national, and international rural landscapes. In the face of increasing breed homogenization, rare farm animals may have become material and symbolic resources through which conventional agricultural and more urbanized rural groupings attempt to negotiate, reassert, and express distinctive regional identities.

The final post-productivist advantage of many native breeds is that they provide an opportunity to promote an ecological approach to livestock production (Alderson 1994), because some indigenous breeds can thrive in habitats that are relatively inhospitable to commercial breeds.[34] Commercial breeds have been developed to fulfill highly specialized functions, and unwanted characteristics have been systematically removed. The animal's productive environment has also been scientifically altered to promote maximum output. All of that is costly to implement. The alternative is to "take the environment as the starting point, selecting breeds that are suited to it" (Alderson 1994, 57). Soay sheep, Tamworth pigs, and Shetland cattle can survive in fairly hostile environments.[35] They can convert relatively unpromising and poor-quality habitats such as seemingly barren shorelines into productive resources, and they require minimal feeding and housing. Such attributes are becoming increasingly valued by cash-strapped commercial farmers who are looking for ways to cut their production costs. This point was raised by one of my rare-breed hobbyists, who observed that the RBST field officer had received a number of inquiries in 1999 from "ordinary" farmers about low-input cattle. She said that such cattle "need very little grazing . . . can stay out all winter, [and] they're not needing [to be] cosseted like a lot of the continental breeds. . . . I think that, with people being much harder up, if they are actually starting up in farming, it's probably the economic way to go." Another hobby contact, who organized pedigree auction sales, observed, "There's not money around farming just now, so these big continental breeds aren't as popular as they were. So [commercial farmers are] looking back again at the [Aberdeen] Angus, the Shorthorns, and the

Herefords and things like that; it's a low input into feeding these animals compared with the big continentals. It's just all fashion."

Furthermore, some rare breeds have contributed to conservation projects aimed at safeguarding environmental habitats, because the animals graze the land differently from non-rare breeds (Yarwood and Evans 1999, 83). For example, because Hebridean sheep forage on purple moor grass as opposed to heather, this method of grazing promotes the regrowth of heather. Evans and Yarwood (2000, 242) also note that the Wiltshire Wildlife Trust uses a "flying flock" of such breeds "to tackle conservation problems at different sites as they are required." Rare breeds are not only "doing their bit" to preserve endangered habitats; in the process, they are also conserving themselves. As long as environmental conservation remains a priority, certain rare breeds will be given a viable opportunity to also secure their own fate.

So far, I have given an overview of the RBST and its role in defining, identifying, and monitoring what it considers to be endangered breeds of British farm animals. This provided a useful context for understanding how rare breeds have become reevaluated and reused in post-productivist agricultural and environmental settings. Although some commercial farmers can see the marketing and conservation advantages associated with rare livestock, my hobby contacts were particularly receptive to keeping these animals. The remainder of this chapter considers why this appears to be the case.

In 1996, two geographers, Nick Evans and Richard Yarwood, mailed a questionnaire to 7,874 members of the RBST and achieved a 23 percent response rate ($N = 1,834$).[36] Most had been members of the trust five to ten years, while a third had joined after 1991. Of those who responded, 57 percent were keepers of rare or minority breeds of livestock.[37] Evans and Yarwood found that keepers were more likely to own sheep (59 percent) than cattle (31 percent), although pigs (25 percent) and poultry (24 percent) were also very popular.[38] The three main reasons offered for joining the RBST were "to help preserve our past," "to support a conservation charity," and "to find out more about traditional livestock." Members were most likely to be retired or already involved in farming. Housewives and professional people were also represented. For example, their sample included veterinary, teaching, and business personnel. My hobby contacts ($N = 12$) tended to be female (75 percent) and married (75 percent); were, on average, forty-six years old; and were employed (67 percent) or self-employed (25 percent).[39] They included two farmers, a gardener, a pedigree-livestock manager, two secretarial workers, a former teacher, a former nurse, a former administrator, two shop assistants, and a research fellow. Three-quarters of my hobbyists were born in Scotland; the rest were from England; and the majority (83 percent) had gained qualifications at further- or higher-education establishments.

Evans and Yarwood (1998, 10) identified three main, but not mutually exclusive, reasons for owning rare breeds.[40] First, 65 percent of keepers did so to preserve rare-livestock numbers. A similar proportion said they kept rare breeds for their produce. Finally, 22 percent regarded and treated their animals as pets. My respondents gave similar reasons. A retired couple who produced both commercial and rare livestock identified conservation as central to their hobby. The husband said, "Some people are brought up with a sense of duty to other people, or in this case, to animals. If you're privileged, in the sense we are, we can afford to do it. We regard it as, 'Somebody's got to do it,' and we've taken it on. So I think there's a duty aspect to it, as well." His wife reinforced this view when she said, "It's very rewarding to build up numbers and know that they're protected for the future, the gene pool that people keep talking about. . . . So it's our duty in this generation to preserve them, for the future." Their use of the word "duty" conjures up a sense of urgency coupled with a moral imperative to save these animals. However, the husband noted that some breeders did not appreciate the conservation angle: "There are still people with wealth who want pretty animals in the park in front." This observation was repeated by another rare breeder, who said that some hobbyists were not interested in "the breeding livestock side of things. . . . [I]f [an animal] never had a calf in its life, it wouldn't matter." Instead, she said, "They're going for something like Highlanders or White Parks . . . something very distinctive [and] nice to look at when you're looking at your nice view." She also described how some people "start collecting animals"—for example, when attending rare breed sales, they will say, "That's pretty. I like that, [or] I'll have that." Although such people are well meaning and "want to save a breed because [it's] endangered, and this is very commendable," they are not "livestock people . . . who have a genuine interest in the breeding and the genetics of farm livestock." The breeder acknowledged that she "used to have pets" and that the "odd one or two [animals would] be [on the] pet side of things. But I don't think everything in the place is a pet. . . . I'm very much a livestock person; I'm interested in other people's livestock, not just my own."

Some of my hobby farmers kept rare breeds for their produce. As noted earlier, rare-breed meat has been promoted as a way to conserve endangered farm animals. Many hobby animals are regarded as "working animals" and are expected, albeit to varying degrees, to be commercially productive. As one spouse of a full-time hobby couple noted, "We had a proviso that any animals that we kept would have to pay their way." They counted on their pigs to "finance themselves," which meant that the couple had to breed from them "to have something to sell." The wife explained that she had wanted to keep rabbits because they were useful, but she wouldn't have to "kill them . . . to make use of them." As both the husband and wife were competent in spinning and weaving, they set up a craft business processing the fiber from their fifteen angora rabbits and twenty angora goats. The wife insisted that part of her "business ethic [was] to try and use everything [the animals] produce. Sometimes it works better than others, but by and large we get rid of nearly everything that they produce as finished mohair items: scarves, shawls, throws, cushions, floor rugs, and that basically takes you through from the youngest to the oldest animals. We stumble along."

Hobby livestock might be expected to earn their keep, but this does not preclude the animals' owners from naming and forming emotional attachments to them. Being responsible for fairly small numbers of farm animals facilitates hobbyists' getting to know their animals as individuals. Even so, they also try to distance themselves from slaughter stock, an issue I return to in Chapter 8. In the main, many hobby animals could be described as productive outdoor pets. For example, a breeder of Jacob sheep acknowledged that her livelihood did not depend on her animals' productivity and that her sheep were, in effect, pets. "I would love a dog, for a start," she said, "but I feel because I'm at work full-time . . . it's not fair to keep an animal in the house all day. . . . [A]t least I know if the sheep are outside, they're doing their natural thing: They're eating grass, and . . . you can get away with just looking at them once, twice a day." She was aware that pet animals typically stay in the house and seemed ambivalent about seeing her sheep as pets:

> I try not to make it sound that my Jacobs are pets, because at the end of the day they're not living in the house with you, whereas some pets . . . usually do. So they're still outside at the end of the day. And I've seen myself in the winter lying in my bed; it's been a howling gale, miserable rain, and you think my poor animals are out in that weather, but there's nothing you can do, whereas a dog or a cat would be lying at your fire and quite cosy. So I suppose I feel there's a distance. . . . [T]hey're not actually in your own surroundings in your own home, whereas pets are.

Her comments raise interesting issues for classification. For example, although horses stay outside or in stables, they can readily be regarded and treated as pets.[41] In contrast, livestock are an ambiguous category mainly because they are a source of food. Those who work closely with livestock can form pet-like relationships with some or all of their animals, but commercial workers seem more hesitant to acknowledge such affection for their livestock. In contrast, hobby farmers openly express their emotional attachment to their animals. The retired nurse said she perceived herself as "soft in the head" because she was not afraid to demonstrate her affection for her livestock. Initially, she compared herself to "society" and then to the "majority." Then she realized that she was comparing herself to commercial farmers, whom she perceives as more emotionally distant from their animals and whom she regarded as setting the norm for how one ought to relate to livestock:

> That little lamb we call Feather, a farmer would have knocked it on the head; it wasn't really viable. It lived, but it shouldn't have, and he would have knocked it on the head. Blind Billy—gorgeous, but he's blind—if he gets into a field by himself, he panics. . . . I think common sense would tell you that it was not a good idea to keep a blind lamb. . . . But if a farmer had a blind lamb, he would take it to the abattoir. I'd have a blind lamb, and I'll keep it and bury it here. That's the difference. And the majority are like the farmers, so it's not society that I'm thinking I'm different from. It's your normal, fairly hard-headed farming community.

Hobby farmers are more likely to describe and relate to their livestock as pets because their livelihoods are also minimally tied up with the productivity of their animals. However, the more a person's income depends on livestock, the less likely he or she will be to allow himself or herself to perceive livestock as pets. For example, the Ryeland sheep breeder was aware that she became more "commercially minded" when her personal circumstances required that she increase her earnings. Similarly, a male hobby farmer explained that, when he started keeping rare breeds, he was "a lot more sentimental" toward them. But he had since renovated his house and perceived more pressure to earn money. It was still "nice to look over the wall . . . at the lambs jumping up and down," he said, but "I don't fiddle about quite so much as I used to. . . . If it's not paying its way, then it's got to go." In other words, animals he had previously kept for sentimental reasons might be sent to market or slaughtered because he needed the money.

Finally, my hobby farmers are not a homogenous cohort. Keeping rare breeds can be the first rung in a hobby farmer's animal career ladder. Some hobbyists may initially acquire rare breeds of livestock because they want to do their bit to conserve such animals or because they want something that is nice to look at in the field. Given that most are new entrants to farming and may have little, if any, animal-husbandry skills, the learning curve can be steep. Hobbyists at this stage in their careers are more likely to perceive, and thus interact with, their animals primarily as outdoor pets. However, the animals may still technically be "working animals" that are supposed to "pay their way." Hobby farmers who become more knowledgeable about and interested in breeding rare breeds are more likely to show their animals. These hobbyists may start to distinguish themselves from, and express criticism about, those whose "hearts are in the right place" but remain relatively ill-prepared to look after livestock "properly." A woman who had kept Jacob sheep for about seven years mapped out her hobby career trajectory from novice keeper to more knowledgeable breeder. She attributed the transition to becoming more actively involved in the breeding and showing side of rare livestock:

> I hadn't really been interested in the breeding side . . . when I joined [the RBST]; I just wanted to have a few sheep. But then as I bought more books on sheep, I started to get more interested in that side. . . . The first year I got a loan of a ram, and then when the lambs came along the following year, that was fine. But I hadn't even . . . thought about showing then. I had heard about doing showing, but it wasn't really until I had been put in touch with [a fellow Jacob breeder] that it really started and got really into the showing.

She appreciated the social side of attending agricultural shows, where she could share her experiences with like-minded people, but tended to be a spectator on the periphery. She then decided to put herself into the limelight. After her first show, she said, she was "hooked on it." With increasing exposure to the showing environment, she identified people whom she idolized: "They've got great sheep, and I'd love my sheep to be as good theirs. . . . I suppose I am sort of trying to achieve now to get good stock." On reflection, she said, "Over these last few years, I've had so much experience that, [in] my younger years, it was just really a case that [the Jacobs] were more like pets. . . . I'll be learning forever, basically, with animals. I think you always are. But I certainly think I know more about the breeds now than I did when I first joined, even though I'd read a lot about it. I think the hands-on aspect and the showing does make you."

Most of the hobby farmers studied were women from non-livestock-farming backgrounds, and they seemed to be the socio-cultural carriers of a more urbanized and feminized understanding of human–livestock relations than is found in the commercial sector. This provides a useful opportunity to compare and contrast two different sets of beliefs, knowledge, and practices. Of course, not all hobbyists are the same, but there seems to be a common progression. Initially, most start as novice livestock keepers. Over time, they become more skilled and confident in handling and managing their animals. Some may go on to develop an interest in, and become knowledgeable about, breeding to the extent of showing their livestock. And a few may aspire to be more commercially motivated. Perhaps this indicates a hobby-animal career. Although most new entrants tend to see their livestock as productive pets, they can rise up through the hobby ranks to become "livestock people" who are interested not only in their own animals but also in those of fellow hobbyists. Even though hobby farming, or part-time farming, has a credibility problem in relation to mainstream full-time farming, it has been described as being "in tune with the times" (Gasson 1988, 7). Good-life enthusiasts may draw attention to their alternative environmental credentials and farm-animal-welfare-friendly approaches, but there is some evidence to suggest that, although they are "well meaning," some may inadvertently harm their animals through ignorance, inexperience, or lack of time. As most of my contacts, many of whom were or had been affiliated with the RBST, were keen to "do their bit" to conserve rare breeds by eating or visually consuming them, moves increasingly are afoot to commercialize such animals. Pursuing a productivist ethic in which endangered livestock have to pay their way, albeit to varying degrees, links into ongoing ideological debates about the extent to which hobby farming might, or might not, be categorized as post- or non-productivist farming. In practice, abstract and dualistic classifications are often found wanting. This case seems no different. Furthermore, although the socioeconomic characteristics of rare-animal hobbyists are fairly typical of those involved in hobby farming, it is possible that their attitudes and experiences toward livestock may not be. In other words, hobby farmers attracted to keeping rare breeds of livestock may be an additional subset of the hobby fraternity.

That most of my respondents were women from cities and towns seemed an important factor in their attitudes toward animals. As city dwellers, they had been less exposed, if they were exposed at all, to the systematic and routine commodification of animals. Yet they were clearly less sentimental and more instrumental toward livestock than some of their urban friends. Generally, their views of their animals lay somewhere between those involved in mainstream farming and those who hold more ethically informed positions toward non-human animals, such as vegetarians. Given the growing trend toward smallholding, this means that a generation of more or less urbanized children is growing up in closer proximity to livestock than their parents did. Hobby-animal farming offers its adults and youngsters an opportunity to reconnect physically and emotionally with farm animals, many of which are destined for their plates. How increased exposure to food animals, albeit on a smaller productive scale, will influence the nature of human–livestock relations remains to be seen. What is more certain, however, is that hobby-animal farming, like commercial livestock production, is built on a number of productive paradoxes. It is these productive paradoxes that I now consider.

6

Sentient Commodities

The Ambiguous Status of Livestock

> You just *canna* financially keep everything that has been a nice
> *wee* cow and follows you about or you'd be bankrupt! So, I mean,
> you just have *tae* . . . part *wi* them. As opposed to if it was a pet
> in the true sense of the word, you'd probably just keep it until
> it died. But although . . . they can be in the broader sense a pet
> because they're friendly *wi ye,* there's no economic reason for
> keeping them if they're no producing. And the same *wi* like a pet
> lamb that you've reared, it's probably one of the most traumatic
> things about having pet lambs that get attached to *ye,* and they
> get attached to *yer* kids and things like that, and trying to get rid *o*
> them, it's difficult. You *widna* sell your family dog just because it
> had something wrong *wi* it, but you *wid wi* a farm animal. (Farm
> stock manager)

Farm-animal workers clearly get to know and become emotionally attached to some of their animals, but the majority of petted livestock still go to market or slaughter. Over the course of the next three chapters, I show that this is just one of a number of tacit productive paradoxes that lie at the heart of producing and killing livestock. I also draw attention to some of the emotional and cognitive challenges underpinning human–livestock interactions, because this provides a useful starting point from which to explore how workers pragmatically manage these challenges. Although these aspects are alluded to in this chapter, I discuss them more fully in Chapters 7 and 8. Here I explore the legal and perceived status of food animals, especially the ambiguous status of stock pets. Clearly, livestock are legally someone's property. Given this fact, I briefly review the changing historical and legal status of livestock in U.K., U.S., and European animal-welfare legislation.

A significant ramification of animal domestication is that animals are turned into property (Salisbury 1994, 15). This is illustrated etymologically by the very term "cattle," which is derived from the Latin word "capital," and "cattle, chattel are the North and Central French forms, of the same word" (Skeat 1910, 97). Moreover, "A chattel mortgage was long considered a cattle mortgage and up until the sixteenth century the English people spoke of 'goods and Cattals' rather than 'goods and chattels'" (Wilfred Funk, quoted in Rifkin 1992, 28). Prior to the passing of any statutory animal-welfare legislation in Britain, it was the role of senior judiciary, in England and Scotland, to adjudicate in their courts of Common Law "the legal status of animals by reference to their standing as property" (Radford 2001, 99–101).[1] The courts classified animals on the basis of ownership. Tamed and domesticated animals were deemed to belong to someone, while wild animals did not. As the jurist William Blackstone noted in the eighteenth century, "In such as are tame and domestic, (as horses, kine [cows], sheep, poultry, and the like) a man may have as absolute a property as in any inanimate beings . . . because these continue perpetually in his occupation, and will not stray from his house or person, unless by accident or fraudulent entitlement, in either of which case the owner does not lose his property" (quoted in Radford 2001, 100).

The courts discerned the financial value of domesticated animals on the basis of how owners used them. For example, working animals used for food and draught were deemed to be of sufficient value to merit legal action if stolen. In contrast, although the theft of pet animals meant their owners might gain some form of compensation, the thief who stole such animals would not be prosecuted. According to Blackstone, animals kept primarily for "pleasure, curiosity, or whim" were treated differently because "their value is not intrinsic, but depending only on the caprice of the owner" (Radford 2001, 101). Except where the maltreatment of an animal lessened its value and was carried out by someone other than the owner, Common Law showed no concern for the welfare of animals. Just as owners could do as they pleased with their inanimate property, so they could do with their animals. There were no legal safeguards governing how owners of livestock ought to rear, market, or slaughter their animals. This state of affairs clearly depicts the extent of men's unbridled dominion over animals, especially domesticated ones, and the lack of direct moral and legal concern for their welfare.[2]

The radical and extraordinary vision of John Lawrence, gentleman farmer and writer, fundamentally challenged indifference to animals' suffering when in 1796 Lawrence advocated that this was an issue of public policy: "I therefore propose that the Rights of Beasts be formally acknowledged by the state, and that a law be framed upon that principle, to guard and protect them from flagrant and wanton cruelty, whether committed by their owners or others" (quoted in Radford, 2001, 3–4). Twenty-six years later, the first state legislation to protect animals was passed in Britain. The Martin's Act provided a degree of protection for certain species of domesticated animals from public and private human maltreatment. The legislation states that "if any Person or Persons shall wantonly and cruelly beat, abuse, or ill-treat any Horse, Mare, Gelding, Mule, Ass, Ox, Cow, Heifer, Steer, Sheep, or other Cattle . . . he, she, or they so convicted shall forfeit and pay any Sum not exceeding Five Pounds, nor less than Ten Shillings, to His Majesty, His Heirs and Successors" (Great Britain 1822, 2–3).

This legal development not only marked the first of many incremental steps to mediate and regulate the ownership and treatment of animals in the United Kingdom; it also turned the maltreatment of animals into a criminal act (Radford 2001, 102).

Just as human cruelty toward working animals has been a long-standing moral issue, concerns about farm animals' welfare can be traced back to the eighteenth century. For instance, Lawrence expressed his displeasure at the anthropocentric and instrumental view that animals were mere resources that had been created for man's purpose. Human dominion over animals had led to their "'natural interests and welfare' having been 'sacrificed to his convenience, his cruelty or his caprice'" (quoted in Radford 2001, 262). In 1911, the Protection of Animals Act, which stipulates, "While man is free to subjugate animals, it is wrong for him to cause them to suffer unnecessarily," was passed (Brambell 1965, 7). This important extension was the belated success of groups that had lobbied since the 1820s to enhance animal protection. However, the impetus for ongoing reform petered out after World War I and would not return until the 1960s (Radford 2001, 89). These decades coincide with public food shortages, poverty during the Depression of the 1930s, and food rationing during and after World War II. People were prioritizing their own welfare during these difficult times: "The forty years after 1918 were a barren period in the evolution of the animal rights ethic; basic speciesism was accepted as common-sense necessity and dissent was dismissed as eccentricity" (Ryder 2000, 143–144).

Central to reviving public and government interest in farm-animal welfare was the publication in 1964 of Ruth Harrison's *Animal Machines,* which offered a seminal critique of factory farming methods. In the same year, the U.K. government responded by setting up the Brambell Committee, which was invited to "examine the conditions in which livestock are kept under systems of intensive husbandry and to advise whether standards ought to be set in the interests of their welfare, and if so what they should be" (Brambell 1965, 1). According to Mike Radford (2001, 262), the committee was to "consider man's responsibilities toward agricultural animals in a new way, an opportunity which it exercised to the full." The term "welfare" also made its legal debut in Britain's legislation to protect animals, but its meaning has still to be "explicitly defined in law" (Radford 2001, 264). Nonetheless, the consternation generated by Harrison's book contributed to the emergence of not only a new interdisciplinary field of scientific study on the welfare of farm animals but a new set of animal experts as well. Such specialist personnel would play a significant role in shaping future policies governing welfare standards for productive animals (Radford 2001; Sandøe et al. 2006).

The Brambell Committee (1965, 13) stated that "an animal should at least have sufficient freedom of movement to be able without difficulty, to turn round, groom itself, get up, lie down and stretch its limbs." These observations formed the basis of a minimal standard known as the "Five Freedoms" and became a moral beacon that would frame national and European discussions (Rollin 2004). John Webster, who was a founding member of the U.K. Farm Animal Welfare Council (FAWC), broadened the scope of the Five Freedoms beyond movement and spatial issues to include aspects such as the animal's diet and health.[3] Although not legally binding, the Five Freedoms, outlined below, provide a point of reference to assist those who formulate, assess, and bear direct responsibility for animal-welfare standards (Farm Animal Welfare Council 2007, 3):[4]

> *Freedom from hunger and thirst,* by ready access to fresh water and
>> diet to maintain health and vigour
> *Freedom from discomfort,* by providing an appropriate environment,
>> including shelter and a comfortable resting area
> *Freedom from pain, injury and disease,* by prevention or rapid diagno-
>> sis and treatment
> *Freedom to express normal behaviour,* by providing sufficient space,
>> proper facilities and company of the animal's own kind
> *Freedom from fear and distress,* by ensuring conditions and treatment
>> which avoid mental suffering

These benchmarks have been adopted by other animal-welfare organizations, such as the Royal Society for Prevention of Cruelty to Animals in the Freedom Food Scheme, which was launched in 1994. The scheme has a dual purpose: to raise standards and to inform and assure consumers of those standards. Producers signed up for this scheme have their animal and dairy products suitably labeled (Royal Society for Prevention of Cruelty to Animals 1999, 1). While the welfare agenda continues to bring about a number of noteworthy changes, there is some concern that the priority of farm animals' welfare might be offset by the declining economic value of intensively produced animals (Winter et al. 1998). For instance, as the mass production of pigs and poultry drives down costs associated with their production and produce, this could also contribute to undermining their welfare. In contrast, niche markets, where customers pay a premium for free-range and welfare-friendly livestock, in principle increase their value and improve the treatment of each animal. Attending more closely to how agricultural policy might affect farm-animal-welfare standards would be one way to discern any unforeseen policy-market consequences (Winter et al. 1998, 321).

Clearly, the moral and legal standing of domestic animals such as cattle, sheep, pigs and poultry has changed over the past century, but the advance has been greatest in Europe. The United States offers little formal protection to farm animals. One of the key U.S. federal laws, the Animal Welfare Act, has been described as "essentially irrelevant" because it does not apply to animals reared for food. To put this comment in perspective, David Wolfson and Mariann Sullivan (2004, 206) state:

> Approximately 9.5 billion animals die annually in food production in the United States. This compares with some 218 million killed by hunters and trappers and in animal shelters, biomedical research, product testing, dissection, and fur farms *combined*. . . . From a statistician's point of view, since farmed animals represent 98 percent of all animals (even including companion animals and animals in zoos and circuses) with whom humans interact in the United States, all animals are farmed animals; the number that are not is statistically insignificant.

The U.S. livestock sector is especially intensive and highly consolidated. This is partially explained by government agricultural policy dating back to the 1950s, which encouraged the development of ever bigger farms (Appleby 2006; Winders and Nibert 2004). The level of institutional power the sector has accumulated over the years has enabled it to secure a pivotal position whereby the industry itself determines whether, and to what extent, its productive practices are problematic. It has been suggested that the farmed-animal industry

> has persuaded legislatures to amend criminal statutes that purport to protect farmed animals from cruelty so that it cannot be prosecuted for any farming practice that the industry itself determines is acceptable, with no limit whatsoever on the pain caused by such practices. As a result, in most of the United States, prosecutors, judges, and juries no longer have the power to determine whether or not farmed animals are treated in an acceptable manner. The industry alone defines the criminality of its own conduct. (Wolfson and Sullivan 2004, 206)

Despite this, a few states have passed legislation. For example, Florida prohibited the use of gestation crates for pregnant pigs in 2002. This legal action is thought to be the first of its kind in the United States to outlaw a specific practice on the grounds of animal cruelty (Mason and Finelli 2006, 121). Two years later, California legislated against the force-feeding of geese for foie gras production (Appleby 2006, 161). As American consumers become better informed about livestock-productive practices and issues relating to animal sentience, this has also generated increasing unease (see, e.g., Pew Commission on Industrial Farm Animal Production 2008; Pollan 2006).[5] Aware of growing concerns, the livestock industry, supermarkets, and fast-food chains have adopted voluntary initiatives to improve welfare standards throughout the supply chain. Much of this attention has been directed toward raising standards for egg-laying hens and animals at the point of slaughter (Wolfson and Sullivan 2004). The McDonald's Corporation, which has approximately 31,000 restaurants in 119 countries, is a significant force in the market (Kenny 2006, 167). In 1999, it approached Temple Grandin, a high-profile animal scientist who has dedicated her career to designing equipment to facilitate the handling of cattle and pigs on farms, feedlots, and slaughterhouses, to audit a number of abattoirs. Initially, these inspections were not taken seriously by the industry, but when a sizeable slaughter plant was dropped from an approved list of suppliers, plant managers began to improve animal welfare and better maintain their equipment. Because McDonald's buys beef from about 90 percent of the large slaughter plants dispersed throughout the United States, such businesses are financially motivated to retain its orders. Grandin (2006b) said that the audits had stimulated the most noteworthy changes she had seen in twenty-five years. Moreover, in recent years, a growing number of small-scale producers have joined independently validated welfare schemes in response to niche-market customers who are prepared to pay a premium for animals produced in more welfare-friendly contexts. For example, the Certified Humane label (which is governed by Humane Farm Animal Care) has established a standard that stipulates: "Livestock must have . . . a diet designed to maintain full health and promote a positive state of well-being" (Appleby 2006, 159–161; see also Fox 1999). Such initiatives can be honored more in theory than in practice, and no amount of emphasis on the humane will satisfy those opposed in principle to the farming of animals. Nonetheless, they are beneficial changes, though the U.S. legislation still lags behind that of Europe (Mason and Finelli 2006; Wolfson and Sullivan 2004).

The legal status of livestock in European law has been fundamentally revised to acknowledge animals' sentient nature. Until the mid-1990s, the status of livestock in the Treaty of Rome was comparable to that of "goods" or "products" (Stevenson 1994, 116). Productive animals did not receive any preferential treatment over inanimate agricultural commodities, such as wheat and potatoes. During the 1980s, animal-welfare groups such as Compassion in World Farming, along with representatives of the U.K. and European parliaments, actively campaigned for the Treaty of Rome to be amended to pay heed to the sentient nature of animals (see Lymbery 1999; Watts 1999). A minor concession was made in 1991, when a declaration was added to the treaty that required agricultural-policy makers in the European Community (EC) to "pay full regard to the welfare requirement of animals" (Radford 2001, 106). Although devoid of any legal standing, it was an encouraging development. Undeterred, the campaign intensified and was rewarded a few years later. Following the Amsterdam Summit, the Protocol on Animal Welfare was ratified in 1999. For the first time, there are "explicit legal obligations to consider animal welfare within the EC Treaty," and it "contains the first reference in EC law to animals as 'sentient beings,' changing their status from mere goods or agricultural products. Member states are now undeniably obliged to protect animals for reasons of morality rather than commerce" (Camm and Bowles 2000, 204). A report by Compassion in World Farming suggests, "If an animal is 'sentient,' it is capable of being aware of its surroundings, of sensations in its own body, including pain and hunger, heat or cold and of emotions related to its sensations. It is aware of what is happening to it and its relations with other animals, including humans" (Turner 2003, 4). The shift from inanimate to animate artifact now overtly acknowledges the sentient characteristics of livestock animals and, in principle, builds in additional safeguards that could enhance their welfare.

However, some argue that despite animals' elevated moral status, allowing them to be owned has the effect of treating them as commodities. Gary Francione (2006, 83), an American professor of law, asserts that the property status of animals "renders meaningless our claim that we reject the status of animals as things." He draws an analogy with legal attempts to treat human slaves more "humanely." The introduction of slave-welfare legislation established a three-tier system: "things, or inanimate property; persons, who were free; and in the middle, depending on your choice of locution, 'quasi-persons' or 'things plus'—the slaves" (Francione 2006, 92). The middle tier, albeit well intended, was the moral equivalent of no-man's land. If human slaves were to be accorded substantial moral consideration, then they "could not be slaves anymore, for the moral universe is limited to only two kinds of beings: persons and things" (Francione 2006).[6] In other words, because slaves were not entitled to equal consideration, they were most likely treated as things.

The notion of "equal consideration" refers to a moral principle whereby "we ought to treat like cases alike unless there is good reason not to do so." Even though humans and animals may not be alike in all ways, what unites both species is their capacity to experience, and thus avoid, pain and suffering. For this reason, people and animals "are similar to each other and different from everything else in the universe that is not sentient" (Francione 2006, 84). By implication, if we are serious in our attempts to proscribe unnecessary suffering by animals, then, according to Francione, "we must give equal consideration to animal interests in not suffering." Despite proclamations to balance the interests of both animals and humans when these interests are at odds with each other, the property status of animals effectively undermines the position and needs of animals "because what we really balance are the interests of property owners against the interests of their animal property" (Francione 2006, 80). Francione believes that the only way to really address animals' interests is to grant them the right to a non-property status. Radford (2001, 103) is not persuaded that the concept of rights holds the answer; not only does it beg the question of which species of animals would be awarded these rights, but also "disputes arising from conflicting rights are as often as not determined by recourse to consequences, even though the outcome may be presented as the application of principle."

Nonetheless, Francione reminds us that domesticated animals are our possessions irrespective of how well we look after them and whether we call ourselves "owners" or "guardians."[7] For instance, guardians retain their right to sell or kill their animals. As Grandin (2006a, 207–208) explains, one of the key legal distinctions between property and non-property is that "I can buy, modify, sell, give away or destroy items that I own." Even though semantic changes do not change the property status of animals, they potentially foster more positive attitudes toward animals. Given that legal spheres require intransigent and unambiguous definitions to operate effectively, this begs the question of how representative these clear-cut statements are in practical contexts.

To assume that the commodity status of all owned animals is fixed in all circumstances does not account for situations in which some animals might be perceived by those working with them as being non-commodities that may or may not be "recommodified" (Kopytoff 1986). Although these perceptions do not legally alter the property status of livestock, they do muddy the neat and static logic of abstract statements. It is precisely these types of experiences and perspectives that shape the diverse, dynamic, and contradictory nature of people's perceptions of and interactions with productive animals in everyday life. In the words of William I. Thomas, "If men define . . . situations as real, they are real in their consequences" (quoted in Janowitz 1966, 301). If agricultural workers regard some of their animals as pets, they will treat them accordingly, irrespective of their legal status. This does not preclude such animals' being treated as productive artifacts. In the end, most of these stock pets will still be slaughtered. Clearly, a fine line separates seeing livestock purely as tools of the trade and livestock as sentient beings. I use the term "sentient commodity" to draw attention to the ambiguous and shifting perceived status of livestock and people's cognitive and emotional attempts to negotiate this fine line in practice. Legalistic and philosophical approaches that seem to bracket off the empirical, attitudinal, and affective elements of interspecies interactions not only tend to underestimate the socio-affective significance of people's experiences, but also provide a somewhat partial and skewed understanding of human–livestock relations. Legally, livestock are indeed things, but in practice commoditization is an ongoing process as opposed to a fixed state. As Igor Kopytoff (1986, 64) notes:

> The production of commodities is also . . . a cultural and cognitive process: commodities must be not only produced materially as things, but also culturally marked as being a certain kind of thing. . . . Moreover, the same thing may be treated as a commodity at one time and not at another. And finally, the same thing may, at the same time, be seen as a commodity by one person and as something else by another. Such shifts and differences in whether and when a thing is a commodity reveal a moral economy that stands behind the objective economy of visible transactions.

Livestock animals are atypical market commodities that have an ambiguous product status. Producers may, to all intents and purposes, routinely regard them as articles of trade, but they also, to varying degrees, have to feed them, clean them out, attend to their health and welfare, and learn how best to handle them. Those who come face-to-face with these animals are acutely aware of their biotic and temperamental attributes and sometimes "get to know" the animals as more than just things (Wilkie 2005, 224). To gain a more incisive appreciation of the perceived status of livestock requires us to consider empirically the diverse and dynamic nature of human–animal productive relations. This not only provides a useful context from which to build up a more discerning and systematic picture of the practical, cognitive, and emotional challenges faced by livestock workers, but it also indicates how agricultural workers negotiate these challenges in the course of their everyday work lives.

To date, we have relied on broad brush strokes to characterize human–livestock relations, but such an approach can mislead. For example, I made the assumption during my interviews that the term "livestock" was unproblematic. However, near the end of my interviews, a couple of respondents forced me to rethink. A mart worker who organized public tours of the mart suggested that people from non-farming backgrounds referred to the animals as "farm animals," as opposed to "livestock." She thought "livestock" was a "technical term" commonly used by those working within the industry. In a similar vein, a female hobbyist associated "livestock" with a "large farm produc-ing animal" such as cattle and sheep that were farmed "on a much larger scale." She preferred the term "agricultural animals" because it "seems to encompass all the little bits and pieces like goats and things. . . . [B]ecause . . . we only have two of them, you don't often hear of people farming goats." She said, "I don't really think of ours as livestock. Maybe . . . it's just a different way of looking at it." Clearly, "livestock" can mean different things to different people, depend-ing on the species of animals being farmed and the scale of commercial or hobby animal production. This not only alerts us to the importance of clarify-ing how members of the public and agricultural workers classify productive food animals, but it also hints at a possible linguistic coding whereby "farm animal" conjures up bucolic images of traditional farming while "livestock" invokes images associated with industrialized animal production.

Although farm animals are commonly referred to as "livestock," the term "implies that we view both food of animal origin and the animals that provide that food as a commodi[ty]" (Webster 1994, 128).[8] It might be worth explor-ing the meanings of such terms more fully in future research, as they may not be interchangeable.[9] As Joan Dunayer (2001, 9) points out, terminology does matter, because "the way we speak about other animals is inseparable from the way we treat them." This view is reinforced by James Serpell (2004, 149), who says that terms such as "food animal" and "productive animal" inevitably constrain people to "thinking about them from an instrumental perspective." Clearly, there is much veracity in these authors' observations. However, in the next section, and in subsequent chapters, I show that such an unyielding statement is open to dispute and messier in practice.

My commercial respondents saw livestock animals primarily as a source of food and a means to earn a livelihood: "They are used for [the] food chain"; "Livestock are there to make a living"; "Livestock production is my livelihood—therefore, money." Unsurprisingly, almost all of my interviewees agreed with the statement that livestock animals are unlike pets.[10] For exam-ple, a veterinarian said, "Livestock animals are kept for breeding or meat—therefore, economics are [a] very important part of these systems. Whereas a pet is kept for pleasure or companionship, it has no set [monetary] value to the owner and doesn't have to grow at [a] certain rate or reproduce." A mart worker explained, "Pets are not bred for eating." And a former slaughter man stated, "I would eat livestock, not a pet." A young female veterinarian who worked in a mixed practice explained how she saw the difference in people's relations with livestock and pets:

> Basically [farmers have] cows so that they can get calves from cows and sell the calves to make money. Or they've got cows for milk so that they can sell the milk to make money. . . . But pets are there because the owners . . . want the pleasure of having a pet—taking it for a walk, having the unconditional love you get from a dog, and all the rest of it. So there is a balance. There is a difference, definitely. But on the livestock side of it, basically, if a cow dies, then they've lost money. . . . [T]hat's the way they'll look at it. So obviously, that's not good, so they get the vet out to try . . . to minimize losses. If they've got a calf that's quite ill and they spend a lot of money on it, then obviously that's . . . reduced the price of the calf, because they've had to pay the vet's bills. . . . If something goes wrong, it's money. Rather than, "Oh no, I've lost my cow" kind of thing, it's more, "Oh no, I've lost some money." Obviously, they do care about the animal, as well, and they don't want it to suffer any more than it would have to, but ultimately it does come down to money, and they've lost money because of something that's happened.

Even though these quotes clearly highlight the instrumental nature of human–livestock relations, all of my interviewees claimed that they enjoyed working with animals, and almost all thought it was important to like animals in order to work with them. The very idea that agricultural workers like animals may seem at odds with what they actually do to livestock. How plausible is it for workers who treat livestock as productive tools that can be mutilated, exploited, and slaughtered to still claim they like animals? The label "animal lover" is typically monopolized by, and associated with, those of us who reside in urban settings whose main experience with animals is through pet ownership, an apparently more positive and benign type of human–animal relationship.[11] However, technicians working with animals in experimental research also maintain that it was precisely "love of animals" that attracted them to their area of work and enhanced their ability to handle the animals (Birke et al. 2007, 106). Lynda Birke and her colleagues also identify a division of emotional labor in laboratory work. Researchers tend to be distanced from the everyday care and routine killing of surplus experimental animals; those tasks are carried out by technicians. Even though laboratory animals are rarely named, technicians are more likely than researchers to identify a few animals for special attention. These animals are regarded as pets and friends and "acquire a quite different moral status from the remaining animals—they are treated as sentient rather than as bearers of data, as a 'living entity rather than as a container housing tissues'" (Arnold Arluke, quoted in Birke et al. 2007, 104). Similarly, those who physically handle livestock, such as stock-people and farmers, are most likely to identify a few animals as "friends," "pets," or "work colleagues."

For example, Paul Hemsworth and Grahame Coleman (1998, 20) note that "humans working closely with farm animals develop relationships with their animals often not dissimilar from those that develop between humans and companion animals."[12] This claim is most likely mediated by contextual factors, such as the scale and type of animal production, the species and breed of livestock being produced, the ratio of stockpeople to animals, and the demographic and personality characteristics of those working with the animals (see, e.g., Seabrook 1994). In my research, 86 percent of interviewees acknowledged that livestock could become pets. But this does not necessarily mean that pet livestock relinquished their commodity status. As a senior auctioneer explained, irrespective of how attached one might become to livestock, "They are still a business at the end of the day. . . . If you have gone down the road of livestock production—be it cattle or be it sheep—that is the commodity at the end of the day that's going to provide the money to go to the supermarket with." One of the most prevalent stock pet experiences concerns young livestock such as lambs and calves. Agricultural workers who rear orphaned lambs tend to become emotionally attached to them and can find it difficult to put them away to market or slaughter. A young mart worker who had been brought up on his grandfather's farm told me that he had bottle-fed lambs that were kept for six to twelve months and then sent for slaughter. "At the time you feel bad," he said, but "you don't lose sleep over it; it's just one of those things." The farm stock manager who opened this chapter stated, "It's probably one of the most traumatic things about having pet lambs that get attached to *ye,* and they get attached to *yer* kids and things like that, and trying to get rid *o* them, it's difficult." A female mart worker echoed these sentiments. She said she tried to avoid treating any livestock as pets because when she had befriended orphan lambs in the past, it "broke my heart when they [went] off to the mart. . . . I want my animals to die of old age, not end up in the butcher's." This contact had a non-farming background but had married a small-scale farmer. Her main duties at the mart were to deal with the paperwork associated with cull, store, and fat cattle. Cull cattle required her to process old dairy cows that were destined for slaughter. These animals had been well handled, she explained. "They'll come up and have their heads rubbed, and that's when I think, uuuh, you know, I don't want this to happen to them. But what can I do?" In comparison, store and fat cattle are young animals, up to thirty months old. They can be large, spirited, and unpredictable.

Before she worked at the mart, the worker said, she had been "accused of making [animals] too human. . . . I suppose I would generally think about animals just as hairier or feathered versions of us." For example, her dog accompanied her everywhere. Some people thought that, because she was childless, she was "babying" the dog and it was a replacement for children. She adamantly denied this: "I don't like children. I wouldn't have one." But she did treat her dog as a "friend," she said, and declared, "I love my animals." She soon realized, however, that working with livestock required her to get tough. "I came [to the mart] with the idea that . . . I was going to be nice to the animals," she said. "I wasn't going to hit them with a stick. . . . I was never going to yell at them. . . . [W]ithin a couple of weeks I realized that if I didn't harden up . . . if I didn't suddenly treat them with a bit more respect and stop trying to pet them, then I was going to get killed."[13] By "respect" she meant that she had to learn to step back from the cattle so they could have their own space.

However, not everyone experienced stock pet relations as traumatic. For example, a mart worker born into a farming background, said that she had felt no unease whatsoever when, as a young girl, she took her pet calf to market. She was aware that some of her friends found this a difficult thing to do, but for her, it was a "joy" to take her calf to the livestock auction "to watch it get sold to see how much money" she would receive. Although she also regarded the young calves and sick lambs as pets, she fully accepted that the animals were there to be used. For her, there was no ambiguity. "It's just part of life," she said. From a very young age she had wanted to work at the mart and thoroughly enjoyed the "buzz of the mart." Early exposure to the mart not only appears to have normalized and, possibly, emotionally desensitized her to the selling of livestock; it had also clearly inspired her to play a central role in the marketing process in the future.

The example of petted stock shows how the explicit commodity status of productive animals can be temporarily suspended to allow a more lay perspective of seeing and dealing with these animals to come to the fore.[14] Orphan lambs and calves can be routinely de-commodified on the provisional understanding that, as productive animals, they will still be sent to market or slaughtered. Of course, the informal process of de-commodifying livestock does not legally alter their property status. But that it happens reminds us of the shifting and ambiguous nature of human–livestock relations. If we simply dismiss these experiences by arguing that the animals remain commodities legally, regardless of how producers perceive them, then we are disregarding some of the practical, cognitive, ethical, and emotional challenges faced by those who work directly with livestock. Partisans on both sides of the farm-animal-welfare and farm-animal-rights arguments might prefer to think that agricultural workers and farmers treat their animals simply as things, but my respondents, involved at various points in the production chain, have rich experiences of personal interaction with animals that blur any simple divide between pets and livestock. Farming is dominated by economic interests; hobby farming is, too, though to a lesser extent. But as the next chapter shows, attitudes other than instrumental rationality can and do come to the fore.

7

Affinities and Aloofness

The Pragmatic Nature of Producer–Livestock Relations

> I certainly couldn't work with a herd where you have no reaction
> from any of them, 'cause it's not the same. No, you need that
> interaction from the cows even if it's just a few. . . . I mean, you
> could be in with a group of people, and if none of them were
> interesting, you'd wonder why you were there. But if there [were]
> a few people in the group that were interesting in different ways,
> it makes it worthwhile. The same goes for the cows in the herd.
> (Agricultural contract worker)

Even though commercial livestock production is overtly governed
by economic interests, non-monetary values can also be discerned.
Hobby farming, by contrast, is less profit-oriented, which allows
more expressive attitudes to come to the fore. The coexistence of instrumental and substantive concerns more accurately depicts the indeterminate nature of the byre face in which commercial and hobby producers
have to negotiate pragmatically their interactions with their animals.
Those responsible for the routine handling of livestock have to come to
grips with seeing some, or all, of their animals as undistinguished tools
of the trade while also regarding some, or all, of these animals as idiosyncratic beings. In this chapter, I suggest that the attitudes, feelings, and
behaviors of byre-face workers cannot be uncoupled from the productive
role of both humans and animals in the practical division of labor (e.g.,
breeding, storing, and finishing) or the socioeconomic context in which
commercial and hobby farming takes place (Wilkie 2005).[1] For instance,
breeders are more likely to feel varying degrees of affinity toward their
animals, while those who fatten livestock for slaughter tend to be more
aloof. The productive career path of livestock (i.e., breeding or slaughter)
seems to be an important factor in shaping the extent to which agricultural workers actively engage with or disengage from their animals. It
is precisely these sorts of commonplace and extraordinary productive
circumstances that illuminate the factors most likely to impinge on a

stockperson's attitudes toward his or her animals. Although people can experience ambiguous thoughts and ambivalent feelings about their role in producing and slaughtering livestock, these themes are explored more fully in the remaining two chapters.

Researchers in France recently explored the nature of livestock breeders' and farm advisers' perceptions of livestock and found three main characterizations: "the animal as machine," the "communicating animal," and the "affective animal" (Dockès and Kling-Eveillard 2006, 243).[2] All respondents emphasized the productive role of livestock, but some focused on the technical side of producing animals, while others tried to factor in the psychological needs of their animals by actively communicating with them. Even though these types of farmer–livestock exchanges did not engender emotional attachments to individual animals, other farmers did experience a "real affection for their animals and practice[d] a kind of empathy with them" (Dockès and Kling-Eveillard 2006, 248). The study also suggests that there may be some correlation between how farmers conceptualize their animals and the type of productive methods they use. For example, technologically oriented producers who subscribe to a more mechanistic perception of livestock were inclined to produce pigs and poultry on an intensive scale. Alternatively, those who see their livestock as sentient beings attributed more importance to communicating with them and tended to breed cattle or pigs and calves for quality-assurance schemes.[3] Finally, farmers who were more emotionally involved with their animals tended to breed cattle, although some produced pigs in outdoor production systems.

The perceptions of my respondents can be divided, too. Some saw their livestock primarily as pets, while the majority regarded them as more or less unspecified "tools of the trade," or a means to earn a livelihood and a source of food. However, this view of livestock as faceless market commodities could, to varying degrees, be moderated by an acknowledgment of and respect for the animal's life world. Even though working animals are ultimately productive animals, their temperaments may also be taken into account, especially by those working most closely with them. A commercial stockman explained: "Out of the cows we have, some of them you could go up to in the field every day and . . . scratch their backs, and they'll stand and obviously enjoy it. Occasionally, you'll see that they'll walk toward you for it." Other animals may not mind being handled, but "once they're out in a field, they'll never come by; they'll never walk toward you. There's other ones that if you try to handle them, they'll do anything to get away. It's just totally dependent on the individual animal." For these reasons, producers can form a range of emotional affinities to and aloofness from their animals. That livestock cannot always be reduced to, or totally defined by, their productive roles facilitates a more multifaceted appreciation of these animals. Such insights, though fleeting at times, can and do occur during fairly routine or exceptional human–livestock interactions of a positive or negative nature. Thus, purely instrumental perceptions of livestock can be unstable and somewhat messy in practice. Animals can be located and relocated along a status continuum that ranges from commodity to companion, whereby the same animal may at times be seen by the same worker, or by a different worker, as a tool of the trade, a work colleague, a friend, or even a pet. Of course, such a classificatory continuum also reflects the full spectrum of people's dominant and affectionate relationships to animals (Tuan 1984).

For example, a commercial stockman responsible for 140 cows and their calves explained that livestock are productive animals, but they also have their own lives: "Animals breathe the same air as we do. They're there for a purpose, but to me, they're more than just bits of beef walking about." To him, the animals were colleagues because "I'm working *wi* them every day. People work with people every day, and they're work colleagues. I work with the animals, not every day, but I see them more or less every day." Of course, he did not perceive every animal as an individual. But when a beast required additional attention, it altered his relationship, and he became "more aware of that animal in day-to-day life." Cows were more likely to stand out for him because they stay on the farm for up to ten years. Store animals, in contrast, were slaughtered before they reached thirty months of age. From his experience, farmers did vary in how they regarded the animals they worked with, but he said that cowmen would "always have a closer relationship" than farmers who finished store animals: "The farmers that buy cattle through the market and keep them for a few months to slaughter them, that is all that animal is to them. It's just a product which they are processing. . . . [B]ut the cows, as I said earlier, . . . could be [there] for ten years. So you are bound to have a different feeling toward them if [they're] there for ten years rather than three months."

He identified having a "different feeling" toward breeding animals. "You get to know different sides of the animal: whether they have a quieter temperament. You'll just get to know the animal better the longer you have it. It's just the same as humans: The longer we know a human, the more [we] get to know them."[4] He also kept a close eye on high-quality beasts by "monitor[ing] their progress" and was acutely aware of the "poorest-quality ones." "You're more ashamed of them," he said, because northeastern Scotland is known for producing "quality cattle. You don't want to be seen producing poor-quality cattle. . . . [I]f there is a poor-quality animal, you don't particularly like other people to see it." This would reflect badly on him as the stockman on the farm. Animals he had no special interest in were simply "there to do a job. . . . They're there to produce a product." Special or extraordinary beasts were "there to produce a product, as well," but they added more interest to his working day: "It would be like working in a factory, you know what I mean? It gives you something different to look at, or it would all turn out a bit boring."

Similarly, I spoke to a farm stock manager who was responsible for a resident herd of about 200 cows. However, because his boss was a cattle dealer who regularly bought and sold livestock, he might oversee up to 1,000 animals. Thus, he managed a fairly constant breeding herd and a transitory population of store animals that might or might not be fattened for slaughter.[5] He thought livestock could become pets, especially orphaned lambs and calves, and went on to illustrate how he saw pet stock in relation to the rest of the beasts:

> They're still livestock because . . . , although they can be friendly and they can be pets, at the end *o* the day they're there for a job. And they're there, . . . in the breeding-cow sense, . . . to produce a calf. If they're not in calf, they have to go for commercial reasons. You just *canna* financially keep everything that has been a nice *wee* cow and follows you about or you'd be bankrupt! So, I mean, you just have *tae* . . . part *wi* them. As opposed to if it was a pet in the true sense of the word, you'd probably just keep it until it died.

Animals he perceived as friendly tended to mean more, and he would look out for them: "You know, if you see them in a field, you'd probably make a detour just to walk past them and speak to them or something." He explained that forming friendships with some of his animals referred to a "sort of compatibility between [me] and them. As opposed *tae,* I mean, you will get some cows that if you get them in a confined space, they're going *tae* go for *ye*. And *tae* me, that is the opposite of friendship." Likewise, walking through groups of cattle, the stockman would also go out of his way to speak to favored animals. "I suppose I do just treat them as human beings, really, because I'm working with them so much." Animals regarded as good at their jobs are also singled out for preferential treatment. For example, a former slaughter worker explained that his father, who had worked on a farm, formed "an element of attachment to certain cows because they were particularly good breeders." Because those animals produced future show champions, they got "special treatment": "They were very well looked after; they got better feed, better penning conditions over the winter, and the best grass fields." He perceived this mutual return between human and animal as a form of "reciprocation."

He was cautious about suggesting that his father had formed an "emotional attachment" to such animals, but he was comfortable with the notion of "financial attachment" in the sense of "You look after the things that look after you." Conversely, animals can come to the fore for negative reasons. As a mart worker brought up on a farm noted, "Of *coorse* [course] the *ither eens* [ones] that *ye* remember are the nasty *eens*. *Ye dinna* forget them, either. I mean, if something's tried *tae* kill *ye, ye dinna* forget *aboot* [about] it. But obviously *ye dinna* form attachment *wi* them."

Overall, stockmen seem to actively befriend a few animals, regardless of the size of a herd. This serves the obvious function of managing the rest of the herd, which makes their job easier, but it also injects interest into their day-to-day work. A young agricultural contractor, who had accrued a wide range of experience by working on different types and sizes of livestock units, suggested that there are always some animals that catch your attention, even in a large herd. This was common on dairy farms but less so in beef-cattle units: Milking animals tend to be more docile because they are regularly handled and more acclimatized to interacting with people.[6] As noted in the opening quotation, he yearned for some "reaction" from and interaction with a few animals because this made his job "more interesting." He likened it to "having someone working with you. It's not a person; it's the cattle, and its sort of your interaction with them is special in a way." In a similar vein, the farm stock manager explained, "Out of 400 cows, you would always have some that were pally":

> If you *dinna* have that, you would just be on a production line. They serve a purpose, as well. . . . If you take a bag of feed out the back, . . . you get ones [whose noses are] right in it. When you want to shift them out of a field, you just produce a bag out of the back of the truck, and they just about knock you over, and you can get them into the next field. So it does serve a purpose. It's part of the reason why you want to . . . have them tame so that they're easier to work with. You don't want to be chasing around the field after them in a truck if you can help it. It's no good for anybody's blood pressure, that kind *o* thing!

Consumer and animal advocacy groups often presume that commercial farmers treat their animals as mere inanimate commodities—hence, the concerted effort by welfare campaigns to raise awareness about the sentient nature of these "products" in the attempt to improve the conditions of their production. Emphasizing that meat is "food with a face," some believe, will enhance the moral status of livestock (Williams 2004, 46).[7] However, sentience and production are not necessarily inimical to each other. Drawing on Michel Foucault's work, Anna Williams (2004, 49) argues, "Sentience can be recruited to assist, rather than impede, production, transforming resistance into compliance." Thus, perhaps somewhat paradoxically, tuning in to and harnessing the sentient and socio-behavioral characteristics of livestock is integral to the smooth running of productive and slaughter contexts. It has been suggested that "adapting to animals" is a facet of the connection that can exist between producers and their animals and requires "accepting their rhythms, understanding their motivations, 'taking the time' and attuning one's work techniques to their behaviour" (Porcher 2006, 64). Of course, the extent to which this is applied in practice might be governed by the scale, species, and intensive nature of the productive context. In this case, it would be fair to say that the industry has no interest in exposing its workers to large numbers of unpredictable cattle; that is both dangerous and costly. Handling animals in a respectful manner enhances their docility and protects the workers' and animals' welfare. It also makes good business sense. In other words, it can save producers, auction markets, and slaughter plants money. "Physical coercion is an individualized means of control, requiring significant labour inputs; bruises and the cortisols produced by stress lower the retail value of the carcass significantly. The industry therefore has had a strong economic incentive to elicit compliance from animals" (Williams 2004, 46–49).[8] For example, a mart manager mentioned that overseas guests who visit the livestock auction to see how it is run regularly comment that mart workers handle animals in a manner that respects their "privacy and what you call a space." The visitors note that "we're very direct," he said. "We know where we want the animal to go, and the animal knows where it has to go. But it's just the animal's instinct to not want to go there." So, the manager suggested, "Instead of us forcing that animal, we just take time and just let the animal go at its own pace." Of course, the auction is the public face of the industry and, as the manager said, "We want to show the public the right way of how to do things."[9]

These accounts begin to illustrate how the same stockman can perceive, interact with, and relate to the same species of livestock in a variety of ways and for a number of reasons. Similar results were found in a European study (Bock et al. 2007).[10] Farmers in the study not only expressed greater affinity toward cows than pigs or poultry; they generally perceived cows as "more likeable"—so much so that, of all the species, cows were most likely to be seen as "friends" or "family members" (Bock et al. 2007, 118; see also Bertenshaw and Rowlinson 2009). Similarly, my contacts, who worked mainly with cattle, were more attached to individual breeding cows than to store animals destined for slaughter. Perhaps this is because the sentient nature of animals is most evident and more readily acknowledged in the early stages of production. As mentioned in Chapter 4, it is more difficult to price a breeding beast. The nearer an animal gets to the end stage, the easier it is to assign an economic value to it. Store and prime animals are easier to reify or perceive as commodities than are breeding cows. But any animal that departs from its routine production role, or stands out for positive or negative reasons from the rest of the herd, may acquire more meaning for those at the byre face and become more than "just an animal." Clearly, the productive career path of livestock is important in terms of determining its economic value. But, as I now show, it also plays a part in the extent to which commercial and hobby producers emotionally engage with or disengage from their animals.

To gain a more nuanced account of human–livestock interactions I suggest that it is important to identify the career path of the animal, from breeding to slaughter, and to locate the point at which the worker's path intersects with that of the animal. Livestock-production contexts can induce structural ambivalence (Wilkie 2005). For example, doctors and other caring professionals practice "detached concern," which require them to alternate between "the instrumental impersonality of detachment and the functional expression of compassionate concern" (Merton 1976, 18). Because a doctor is incapable of expressing simultaneously the dual requirements of his or her role, Robert Merton (1976, 8) suggests, people who face this predicament experience ambivalence, "not because of their idiosyncratic history or their distinctive personality but because the ambivalence is inherent in the social positions [roles] they occupy." A similar tension can arise for those at the byre face, given their contradictory productive roles as animal carers and economic producers of sentient products. The nature of this tension appears to be mediated in some measure by (1) the stockperson's and livestock animal's position inside or outside the commercial production process and (2) whether the producer perceives and interacts with the animals on an individuated basis or as part of a faceless herd. Each stage of the livestock-production process (e.g., breeding, storing, and finishing) not only gives agricultural workers a number of routine and unexpected opportunities to interact with and disengage from their animals, but it can also set limits on the frequency and quality of these interspecies liaisons.[11]

For example, a hobby farmer asserted, "I would expect most livestock farmers, and most of my neighbors are livestock farmers, to have some degree of sympathy and affection for at least some of their animals. . . . [F]riendships do occur between man and his livestock: a favourite ewe and so on." Of course, pragmatic reasons might underpin such affection: "I also think that you'll get the best out of an animal that you regard as something above a walking larder or money on legs. If it's just a commodity, I think it becomes easier just to treat it as such." The scale and type of animal production significantly mediates the extent to which producers can realistically engage with their animals, and he acknowledged that it is difficult when a farmer has fifty cattle in a field:

> He's bought them in as calves; they are just cattle. Nevertheless, [there's] the proprietorial eye that looks at them once or twice a day, that checks for problems, that makes sure that they're wormed regularly if necessary, that makes sure that there's no injuries or what have you. I see it all around me, and that to me is how farmers have always behaved. A certain involvement, but just at arm's length, 'cause after all, on the scale that they're working on, closer ties are more difficult.

He considered how he dealt with his own stock: "It's impossible not to have some kind of attachment to a dairy goat, because you're working with her, and she's helping you. She's feeding you." The frequency and regularity of milking had fostered an affinity, or a "sort of closeness," to the animal. "The more you have to do with an animal on a regular basis, the more difficult it becomes to just treat [it] as a commodity," the hobby farmer said. "The animals that have the higher intellectual capability, I think, tend to be the ones with whom we have the closest rapport and who we regard in a slightly different light." This applied to his pigs and dogs. Keeping animals helps one appreciate that each animal is an individual, with its own personality and temperament. For instance, he said, "All those lovely little fluffy bunnies are all pretty much like peas in a pod, and yet they have character differences which enable some of them to worm their way into your affections and become favorites. Each of those creatures is subtly different from its neighbors, and the more you have to do with them, the more you notice this. . . . [E]ven in a flock of hundreds of sheep, . . . they are different." His wife agreed: "That's quite a chilling thought, actually, when you realise it. . . . But if you took the trouble to get to know all the sheep that were going to go off to the abattoir, then you would realise that they were all different. . . . [Then] we'd all be vegetarians, 'cause [we]'d realise they were all individuals." This insight had encouraged her to rethink her perception of sheep. They were not "little robots wandering around" but "feeling creatures."

Another hobby farmer explained that breeding stock is more likely to be named than store animals:

> Pedigree stock all have pedigree names. Some of them are very long-winded, so they're shortened to pet names. . . . [T]hat's OK, because it's going to be here for life. The progeny from the breeding stock

don't get named at all; they're just pigs, and the collective "pigs" covers them all. If one is singled out to replace the breeding stock and it's registered, . . . then it has its pedigree name and gets its pet name.

Once he had earmarked the replacement gilt, the rest became just "pigs."[12] This productive decision changed his behavior toward the animal: "When I was casting an eye on the pigs, I was looking to see that she was growing correctly [with] good conformation. I was looking to see that they were growing fast. So there was a slight difference, and . . . I would make a point of petting her. . . . She's not just a pig anymore; she's come a stage above that, maybe— certainly not treated quite the same as the things that were destined for the abattoir. They were viewed with an eye to get the finish on them as quickly and as sensibly as possible." He acknowledged that, if the pig had not been destined for breeding, "She would be viewed with the same eye as the rest. I think the only distinction is that I have marked her to not go to the abattoir and carry on breeding in that line."

These quotes draw attention to how the productive function of livestock affects how they can be perceived. In this case, a breeding animal became more than just a pig and was treated differently from the animals heading for slaughter. Reproductive animals have a longer life span, are more likely to be individually recognized, and enjoy an elevated status. In comparison, slaughter animals are predetermined to have a truncated life, they tend to be lumped into a de-individualized collective category (i.e., pigs), and they are readily commodified. A senior auctioneer similarly said that commercial breeders "would have an extremely close relationship with their animals"; even though some finishers might experience a similar relationship, in the main their animals were "tools of the trade." Since finishers process a transient population, they tend to see their animals as "just something to buy, fatten, and sell, and [they are] probably . . . motivated by profit."

Finishers have been likened to businessmen, rather than farmers, because, as the former slaughterer noted, that part of the process requires much less investment than the breeding end: Breeders have to invest a lot more time and money and require the services of a bull, which can cost thousands of pounds. If they are going to grow their own feed to winter animals, they need access to machinery such as combines, tractors, and bailers. Many finishers require only access to grass to keep animals over the summer months because they sell them once they are fattened up and in their prime. Having said this, some traditional small-scale finishers are critical of how commercially run units treat their store animals. For example, a mature mart worker who worked on a family farm until the late 1980s compared how animals used to be handled with how "farmers nowadays" perform these tasks. Her family used to walk among the beasts in the field, she said, whereas these days "the majority are just in *wi* a tractor or a jeep and *hiv* [have] a look, flying around and *oot* again. They don't spend time *waaking* [walking] the field. They *niver* [never] see a human." She suggested that, since farmers have less interaction with their animals, the animals can be more wild and difficult to handle when they come into the mart. She said that such farmers are more inclined to see their beasts as "just a means of an income. . . . As *lang* [long] as they're *stannin* [standing] on their four feet and *waaking,* they're OK." When I asked her how she saw the animals, she said, "They're a living thing. . . . An animal will respond *tae* something *spikin* [speaking] *tae* them, definitely. And I think farmers nowadays, they have more acres, more animals, but *hinna* [have not] the time. We *hid* [had] the time; we took the time." But since "everything's more commercialized, cattle are *jist* a means to a living now; the quicker *ye* get them fed an' quick all over." Similarly, another mart worker in his sixties thought farmers were "lazy *noo* [now]" because "they *jist ging* [go] through *wi* a Land Rover. Besides, if they walked through them, the beasts know *ye* and come *till* [to] *ye* like."

The senior auctioneer also highlighted the personal reward breeders experience if they intervene in a problematic birth and then witness the newly born animal successfully suckle its mother. A commercial farmer who had 250 breeding cows and finished most of their offspring explained why she prefers the breeding stage of the productive process: "There is nothing more satisfying than spotting a cow calving, watching that calving progress to the point of realizing that she's in trouble, getting the cow inside and helping the calf [be] born alive. It just never leaves you, the excitement of getting a living, healthy born calf" (see Fukuda 1996). In contrast, the "finishing side [is] not as exciting because they're born, and not so many things can go wrong with them by that stage. February, March, and April are such busy months here, we calve maybe 160 to 180 cows in those three months. The finishing side, well, I mean, they just eat grass, and then every month the fat man comes around and picks out fat ones and that's that, end of story."

The animal's productive function not only shaped the types of relationships this commercial farmer was likely to form with her cattle. It also set important time parameters on that affiliation. Clearly, this influenced the intensity and durability of these associations:

> I would like to think that I would be able to tell you where [a cow] was when she calved last. I'd like to think that I would be able to know that cow and know her history, and know her past, and know if there had been a problem. . . . There's a longevity of relationship there for

maybe ten, twelve, fifteen years. But with a finished animal, you have thirty months to finish it. You can't wait any longer, because then she goes into the skip. [If] she doesn't, well, she'll leave you £300-odd instead of leaving you £500 or £600. So you have a thirty-month window in which that animal is with you, no longer, whereas a cow could be with me for fifteen years if she's lucky.

A mart worker reiterated the importance of time in explaining how he and his father saw breeding and prime sheep on their farm. "The fattening lambs, if *ye* buy them as stores, *ye* just have them a year, and they're sold as fat. You don't really have any time *wi* them; it's . . . all business. Whereas the longer you have an animal, the more *ye* work with it; *ye* get to know it." His father, the mart worker said, usually names his breeding ewes and the animals he has had for a longer period of time because he really "gets *tae* know them." Those animals are named "Blackie" or "Old Ewe," he explained. "It's quite often one or two that don't go away when the rest get sold. . . . [H]e *kin' o* gets attached *tae* them." However, if farmers are forced to interact with any animal, regardless of its career path, that animal is more likely to stand out. For instance, as the female commercial farmer explained, if a newly born animal is "100 percent," she tends to forget about it. However, if she has to handle it, she will remember it. This also applies to "snags," or animals that require her attention because they are ill or require some form of physical intervention or assistance:

> Behind my house I have a field of store cattle which are just about to be finished, and there's four animals in that field that I can tell you where they've been all their lives, from the day they were born to right now, because they had a major problem during their life that I had to help them through. So, yes, that's important. Then there's the rest of them. I couldn't really tell you who their mothers were or where they came from or whether they were born here or at [another] farm. I'm not interested in them because I haven't had to help them. But there's four out there that I know quite well, because I've sat up with them in the middle of the night or I've had to physically feed them or I've had to inflict quite a bit of pain on them so that they [could] get better.

A mart worker who was brought up on a farm also said she was more likely to get attached to an ill animal: "If *ye've* [you've] *hid* one *nae weel* [well] and *ye've hid tae* spend time *wi* it; obviously you're *gaan tae* get *tae ken* [know] it better if *yer* spending time *wi* it." As a result, she "treated it *mair* [more] individually as opposed *tae jist* one *o* the beasts." A commercial organic vegetable farmer who produced rare breeds also identified with animals he had seen through hard times because they tended to "mean a bit more" to him, especially if they were home-bred. Although in theory a farmer should have the same kind of feeling for all of the animals, he said, he was aware that this did not work out in practice. This indicates how producers who rear and breed their own animals might form more personal biographies with such animals. Of course, long-standing bloodlines of livestock are passed down from previous generations. As Bettina Bock and colleagues (2007, 119) note, "This provides a connection between the history of the animals, the farm and the family. Farmers do not have such a sense of continuity with animals reared for slaughter and if the fattening animals were bought instead of bred on-farm there is no historical link at all."

Based on observations derived from large animal practice, a junior veterinarian said that breeding farmers were "more aware of the animal's needs than someone who's finishing." Because more can go wrong during the breeding and birthing process, breeders generally have specialized knowledge and skills to deal with pregnant animals. They also spend much more time with birthing animals, which increases the opportunities to get to know them. At the finishing stage, the vet said, "They're just feeding them, bedding them and mucking them, and looking out for sick ones, and that takes a lot less input than the calving side of it." Unless store animals deviate from the routine fattening-up process, they require minimal intervention. Different productive stages need varying levels of human intervention, which influences the extent to which byre-face workers are likely to identify with or become fond of some of their animals. As the vet noted, a farmer who calved a cow would not only recognize the animal but also recall, for instance, that she had "had twins last year, she was mad last year; she beat everyone up. [Breeders will] know [the animals], and they'll be more in tune with them." Those who fatten store animals, by contrast, "would be minimally attached to their beasts because it's just something that eats a twentieth of whatever the batch of twenty eats." In other words, a store animal being finished for slaughter "doesn't count as an individual at all."

To illustrate the aloofness of farmers toward animals designated for slaughter, I refer to the experiences of a middle-aged couple who started farming in the early 1980s, when they relocated to Scotland from the south of England. They breed and finish Aberdeen Angus cattle and have a herd of thirty to ninety animals. The husband is the main stockman; his wife helps out on occasion, as does his son. The son studied agriculture and has worked in a variety of agricultural contexts. He is thus well placed to draw on experiences that go beyond those gained from working with his father. To provide a context, I start with the wife's observation about slaughtering livestock:

What I find most disturbing, actually, is not killing the animals that are designed for slaughter; it's having to slaughter old breeding animals that have been with us for years and years, and they've fallen ill or something, and I would love to be [like] rich farmers who could just shoot them out the back. I find that very hard—sending these old girls down the road. That's much, much harder for me than the young animals that were designed for slaughtering.

The husband shared this perspective and added: "I suppose the young ones you do tend to hold back from, because you know that . . . their lives here are limited." The practical implication of holding back was that "you don't go out of your way to make friends with them. You restrict that. That doesn't mean that you don't. There are characters, but fewer of them." He explained that, of his current group of twenty-eight yearlings, he knew about five of them. "The rest tend to be a herd," he said. This led the man and his son to acknowledge that they felt "guilty" about taking animals to the slaughterhouse. The son explained, "[I] only feel guilty for ones that are petted. If they're not petted, it just doesn't bother [me] at all." He viewed the non-petted livestock as "just animals. . . . [I]t's not that they're their valued any less, but in a bad way, I suppose, they're sort of a commodity rather than an animal, if you want to look at it from the harsh way." The male animals will definitely be slaughtered, he said, but he had more room to maneuver with the females because he could "twist dad's arm" to keep the odd one as a breeding cow. The petted stock were "more humanized," while the "other ones are more like a deer in a field; they're just there for food. They're not there for any other reason." Later in the interview, I asked the father how his son's notion of "just animals" related to his earlier point of "holding back": "I suppose, with the cows you actually make friends with them, you actually go out and some sort of bonding, some sort of relationship, gets set up. You recognize characteristics, you recognize the fact that they're different, you recognize their uniqueness. With the young stock, you don't do that actively; you do it passively. [S]omething [may be] thrown up in front of you, . . . but you don't actually seek it."

The couple agreed that the cows were pivotal to producing the herd's future progeny and keeping the business going: "You expect them to be here for ten years; the other ones, you don't. Yes, and they're going to be the mothers; they are the sequel, if you like." However, it was precisely because breeding cows reside on the farm for many years that the couple could defer the decision about killing them. "You actually . . . put that off, so you forget that you'll ever have to do it." But, the husband said, "It's going to be harder when [the time] comes. It's probably emotionally easier to buy in calves and slaughter them year in, year out."

These quotes illustrate how the animal's career path influences the extent to which byre-face workers actively engage with or disassociate from livestock. A recurring theme is that store animals are basically perceived as meat on the hoof. Unless they deviate from the routine fattening-up process, they are processed as just one of the herd and will live out their abridged and uneventful lives relatively unnoticed. In other words, slaughter stock is more easily denied the individuated status imputed to and enjoyed by many breeding animals. Just as the farmer's son likens store animals to a field of deer, Kenneth Shapiro (1989, 184) writes, "'The deer' refers to a species as a reified entity rather than as an aggregate of individual deer. When a hunter kills one of the deer, that particular deer is not actually acknowledged as such, but is considered as part of the reified notion—'the deer.'"

Finishing cattle is a productive context whereby "the beasts," in keeping with the local collective vernacular, have become the focus of producers' attention while the individual animals making up the herd recede to the point of non-existence: "What is there is an impersonalized, deindividualized but reified abstraction. . . . [T]his reification of the species dissolves the individual deer and invests the aggregate of, now, non-individuals with a kind of unified being that allows members of the species to be killed as if they were so much grass being mowed. In Heidegger's term, this deer is lived toward as a 'standing reserve,' a resource there waiting for our use" (Shapiro 1989, 185).

Moreover, as each animal is systematically slaughtered, it is simply replaced by another one of its kind—that is, because the animals are perceived as interchangeable, they are easily substituted. The next animal waiting in line simply assumes the recently vacated "serial identity" that is made possible "by the fact that each entity is lived toward as a receptacle of certain features common to the group" (Shapiro 1989, 186). The shared attribute most valued in store animals is their ability to rapidly accrue weight to become prime animals standing by for slaughter. The transient and insignificant nature of their lives and deaths is captured by the following comments made by respondents:

I mean, you're not going to get so attached to them. They're more of a short-term means to an end. Yes, you look after them. Yes, you take care of them—feeding them and housing them—but only because the more care you take of them, the better they're going to fatten up, [and] the more money you'll get for them at the end of the day. (Former slaughter man)

Your store cattle and your feeding cattle are kept in bigger parks [and you] have them for much shorter periods. I mean, it's just a period o months really that you're speaking [about], and you're just shoveling it in one end and out another, and they're going away on a float and they're forgotten about, and the next ones [are] on the ground. (Farm stock manager)

Prime animals are "just animals" or "tools of the trade." They are the finished commodity. Or, as one mart worker put it, "Well, that's *fit* they're bred for, they're bred for beef." A key responsibility of finishers is to ensure that cattle thrive and gain weight. The term "prime" denotes being "first in quality or value, first-rate . . . the time when a thing is at its best" (Collins 1993, 905). When young people are killed in car accidents, it is often said they were "in their prime" or "in the prime of life"; their shortened lives and premature deaths cause much distress to grieving relatives. In contrast, finished animals *have to be* in their prime to be routinely killed for human consumption. This takes us to a central paradox of food-animal production: the cultivation of animal health for the purpose of death. As most byre-face workers enjoy working with livestock, this obvious fact can be a tricky one for them to negotiate. For example, a commercial farmer explained:

I've got four [prime animals] going away tomorrow morning from the field here behind the house. I've been feeding them in the mornings, and I know them all; I can stroke some of them, keep them nice and quiet, and I'll take them in, put them in a pen. I'll wail out the four that I need, load them up, take them up to the killing house, and put them in the lairage there. But I don't want to see them going down the chute and actually having the bullet put in their face—don't want to see that, hear that, or know about that. As far as I'm concerned, they left me healthy, and I'm looking for that check the next day. But I could not see them being shot. That's not my job.

She acknowledged that seeing livestock being slaughtered was a difficult experience: "I didn't want to see a fit healthy animal walking up that ramp and being shot. I didn't want to see that. No, I turned away. I remember that." As I vividly illustrate in the next chapter, such sentiments are not unusual. Although commercial producers, auctioneers, mart workers, veterinarians, and slaughterers are integral to rearing, marketing, treating, and killing food animals, this does not mean they are unperturbed by the slaughtering stage of the production process.

Furthermore, even though recreational farmers keep less livestock, they also try to remain emotionally detached from slaughter animals. A hobby farmer said, "I try never to get to know the ones that are going to be slaughtered as individual animals because I don't like the idea that they're eaten."

However, keeping her distance from "commercial" animals can be easier said than done. For instance, she had a big black lamb called Bruno who was so friendly, she said, that "he'll probably never go." She acknowledged that naming the lamb was a "mistake"—which probably means Bruno will not be sent to slaughter. A rare breeder who does eat her livestock said she accepts that "things have to be killed" and explained that she rarely handles those "destined for slaughter" because this enables her to dissociate from them. Most people make a conscious decision to disconnect from meat animals. Interestingly, a small-scale sheep farmer put a slightly different slant on the issue, explaining that, once there is no further need for frequent and intensive interactions with her ewes during the lambing season, she put the animals outside and continued to check them every day. Despite attending to them regularly, she felt that over time she grew "away from them." However, she also noted that the animals contributed to the distancing process by "tak[ing] off" when she tried to approach them. As a result, she said, "They make it easier . . . because they break that friendship."

Of course, those relieved of day-to-day husbandry responsibilities are less likely to form attachments to livestock. For example, a farm estate manager affirmed that stockmen "undoubtedly . . . get attached to the animals," but since he was no longer around them, he did not: "I'm not looking at them often enough. I'm much more at arm's length. So I don't . . . know them as well as [the stockmen] do. They know them far better because they're seeing them every day." Similarly, a mart manager who had dealt with cattle on the family farm now had little direct involvement with animals at home or at work, which changed how he perceived and related to livestock. When he worked on the farm, the beasts were "an opportunity to make money." He fed them every morning and night. "They were part of the family; they were part of the holding, part of the farm." (He clarified this by noting that he did not really mean they were part of the family and suggested the term "regularity" instead). This meant that "you knew them individually; you knew which one would come first to the bag, the pecking order and such like. You knew which two would always stay back, and then they would come in at a later stage." Even though he no longer fed the animals, because other men were employed to do that, he did pick out animals ready for slaughter. However, he said, "I'm not near so close to them as what I used to be. Now . . . it's more a part of my job." Although he had given little thought to his reduced contact with the animals, on reflection he realized, "I do miss feeding, I do miss spending that time with them—even husbandry, you know: injections and general care of the animals. . . . [T]he personal side of it's totally disappeared now."

However, he also recounted that, while helping his brother to clip some cattle for killing during a visit, his sister-in-law asked him whether he recognized a particular beast. Initially, he said no, but then realized that he did and asked, "Is that *nae* Susie's calf?" The calf, he explained, had the same type of face as its mother: "That was the first cow in calf that my *brither* [brother] and them ever had, and it's *laisted* [lasted] for four years now, that cow. So it's doing well for itself. It's producing good calves." In contrast, even though he saw vast numbers of cattle on a daily basis at the mart, "They just don't mean nothing to me now; they're just part of my job." He described how he used to "walk past cattle in the lairage area and stop and study them" to consider their potential. Now he is too busy. When he surveyed the vast array of pens in the mart, he was still looking at the animals, but from a different perspective. He sought out injured animals or any animal that might be uncomfortable because the pens had been overstocked. By prioritizing the business and animal-welfare side, his focus of interest had shifted away from assessing the potential of the beast and "whether it would be capable of leaving you money."

In this chapter, I illustrate the features of animals and working environments that have helped to shape the views of my respondents. The commercial production of livestock is like any other business concern in that it is driven and constrained by financial imperatives. However, farm animals are an atypical market commodity because their sentient and social natures can make them unpredictable. Livestock animals are not all the same; nor are the people who look after them. To appreciate more fully how byre-face workers make sense of the different ties they form with their animals, I have focused on the career path of the animal, from breeding to slaughter, and the point at which the workers' path intersects with that of the animals. I have suggested that breeders require more knowledge and skill because more can go wrong during the breeding and birthing processes. The close physical contact between farmers and breeding animals and their offspring permits and encourages empathy. Producers' handling of animals—if, for instance, they become ill—increases the likelihood of such animals' being recognized as individuals, and some of them may be named. Those animals, especially if home-bred, can become personally meaningful to the extent of acquiring biographies and breeding histories. As the workers get to know these animals, they form varying degrees of emotional attachment to them. These comments could equally apply to those working with bulls or showing livestock. As long as breeding animals continue to be productive, they remain on the same farm for years; these particular human–animal relationships endure. In contrast, store animals can change hands a few times before they are slaughtered. To comply with BSE regulations, the life span of beef cattle is no more than thirty months. There is little opportunity or need to practice empathetic stockmanship because beef cattle have a short life, are transient, and require minimal handling after the calf stage. Unless they digress from the normal process of production, slaughter animals are fairly anonymous and will be processed as part of a de-individualized and commodified group. Even though slaughter is a pivotal part of the production process, many of those involved in rearing, marketing, treating, and slaughtering livestock try to disassociate from this stage. The division of labor partly facilitates this distance, but there are times that producers, who more readily identify themselves with livestock, do come into contact with deadstock, too. Clearly, slaughter is the final and inevitable stage of meat production. It is the transition from animal life to animal death to which I now turn.

8

Livestock/Deadstock

Managing the Transition from Life to Death

> The final end of a farm animal is probably quite unceremonious, really. You know, one day you just ring up and say, "Your end's come," and although you might not particularly like doing it, you know, you just do it. Whereas say the end of a dog, it is more you know you'd have to consult everybody and you know it would all be done as nicely as possible for everybody concerned. (Organic dairy farmer)

Animal slaughter is an inevitable part of producing meat: One minute you have livestock, and the next you have deadstock. It is a pivotal transitional stage in which "every animal must be killed by bleeding, and this must be done in an abattoir; these two conditions must be fulfilled for the meat to be deemed suitable for human consumption" (Vialles 1994, 33).[1] Some present this transition as a relatively clear-cut event. A former slaughterer initially explained, "I can't say that I ever thought of myself as being surrounded by death. You were surrounded by cows or surrounded by sirloin steaks or whatever. No, I can't think of death as such really come into it." For him, death was an inappropriate term to describe the loss of animal life. "You associate death with people and bereavement and mourning things like that, which are not descriptions or words or emotions that would apply to the abattoir situation, unless you're a card-carrying vegetarian and you can afford to view the animals as having the same . . . feelings and emotions as humans." Indeed, many vegetarians, and all vegans, do object to killing animals and abstain from eating "food with a face." Also, consumers divorced from the day-to-day realities of producing and slaughtering livestock can easily dissociate from the animated source of meat on their plates. But how do people who actually rear, store, finish, market, medically treat, and slaughter livestock make sense of, and feel about, the killing stage? To what extent do they perceive it as an agricultural fact of life or a problematic phase that needs to be managed? My findings indicate that livestock workers can be troubled about and critical of what happens at the abattoir and those who work there. For instance, most byre-face workers freely admit they could not routinely kill animals, especially healthy animals—which, of course, livestock have to be to enter the food chain. Furthermore, not all slaughter workers can stomach killing beasts, either; once an animal is dead, some are more able to disassemble it. Although all stages of the production process throw up expected and unexpected paradoxical challenges, this is accentuated at the point of animal death. It is the nature of such challenges and how agricultural workers respond to them that this chapter explores.

Routine Slaughter: Just Part of the Process?

A former farmer justified sending his animals to slaughter by saying, "I was *aye* needing the money." As one mart worker put it, "Livestock is deadstock."[2] Food animals are standing reserves of meat and a source of income. To perceive livestock as just tools of the trade, such beasts "must be deanthropomorphized, becoming lesser beings or objects that think few thoughts, feel only the most primitive emotions, and experience little pain" (Arluke and Sanders 1996, 173). Some animals may temporarily transcend their ascribed tool-like status, but as a rule, most food animals, whether favored or not, have a common fate: the slaughterhouse. When that day comes, it is quite often byre-face workers who have cared for the animals who will send them to their death. Given the contradictory nature of their role, many workers can experience varying degrees of structurally induced ambivalence (Merton 1976). Workers facing such ambivalence are characterized by overstated or fluctuating actions whereby they suppress one aspect of their role while focusing on the other, or they move backward and forward between the dual functional requirements of their role (Weigert 1991, 42). The ambiguous commodity status of livestock can further complicate matters. Because they are sentient commodities, producers can be unsure how to perceive them and uncertain how to feel toward them. Of course, indistinct perceptions and ambivalent emotions lie at the very heart of interspecies interactions and are intimately bound up with what has been called the "'constant paradox'—the definition and treatment of animals as functional objects, on one hand, and sentient individuals, on the other" (Sanders 2006, 148). This partly explains why byre-face workers, especially those who have the power to decide whether an animal lives or dies, have to learn how to calibrate their thoughts and feelings toward the animals they work with. For example, one hobby farmer said she enjoys saving animals' lives but was "not so good at playing God at the dead end." She is not alone.

A former commercial farmer explained why he tried to remain detached from his animals: "I think your judgment could become slightly clouded . . . as to when you should be selling it, or when you should be putting it down the road [for slaughter]. I think it is purely a thing just to keep your commercial mind free of any sentiment." However, he also said that he had become attached to a twelve-year-old cow; although he intended to sell her at the mart, he ended up taking her home. "My excuse was that the price offered wasn't good enough. So I think, I could justify, but having said that, I mean she's twelve years old now, so she's getting on a bit, and perhaps a more commercializing man [might have] let her go." He reiterated the importance of "commercial realism": "If you did get attached to too many of your animals, then that commercial aspect would go out the window to a certain extent." To foster aloofness, he said, he tries to remind himself that the animals are "just passing through. . . . There is no permanence about it." Although prime animals are destined for slaughter, an experienced farm stock manager explained, "It's the part I don't like. You *canna* afford *tae* think about it too much. . . . When they go away in a lorry, that's them going away, and there's another lot coming on, and I'm doing my job." He used the phrase "going away" as a euphemism for going to the slaughterhouse. He was also adamant that he would rather "sweep the streets or empty dustbins than see animals getting shot and skinned and everything." He said he disliked the men who worked in "places like that" and perceived many of them as "empty heads" who headed for the pub as soon as they put down their tools. A drover at the mart said that he perceived slaughtering as a "dirtier job than working *wi* animals *doon* [down] in the byre. . . . I know somebody that works in taking out the intestines . . . in a killing *hoose* [house]. I just couldn't do that job." The farm stock manager reckoned that "nobody likes to see death; nobody likes to see animals dying." He also speculated that, even though slaughter workers have minimal contact with live animals, this would not preclude them from admiring some of them, especially "really good animals, and for them just to be going in and getting a bullet . . . no I *cudna* [could not] cope *wi* it."

These accounts begin to highlight an uneasy acceptance of what must happen to the animals while clearly spelling out what each worker will, and will not do, in terms of the production process. Interestingly, workers who experiment on animals negotiate similar issues. Lynda Birke and her colleagues (2007, 91–92) found that scientists and technicians "tended to present themselves through a kind of rational emotionality—that is, they defended animal use, including their own, but acknowledged past and present emotional reactions to it in the form of admitting that there were some things they could not do." Such workers clearly get on with the job at hand, though doing so requires varying degrees of "emotional work" (Hochschild 1983).[3] This applies to byre-face workers, as well.

For example, a mature mart worker who once worked on a family farm fully accepted that animals are slaughtered: "It's *jist* something that *ye've* been brought up *tae. Ye jist* know that the animal has to go, it has *tae* be killed." Despite years of exposure to this fact, she still had to prime herself emotionally: "*Ye* know at the end that you're going *tae* sell them, and I think you *jist* sort [of] steel yourself for the idea. They're *gaan tae* be sold, but it *dis* [does] bring a lump to your throat, I will admit that." She explained: If somebody said, "'That beast, we're *gaan tae pit* [put] it *tae* the *slauchter-hoose* [slaughterhouse], kill it, and we're *gaan tae* bring it back *tae* [your] freezer.' . . . *Ye hiv tae* have a hard nerve a lot of times. *Tae* think that beast *wis* [was] *waaking* on four feet a week ago, now I'm eating it. *Ken ye jist* forget the emotion side of it?" Of course, overriding one's emotions can be easier said than done. An experienced mart worker who has worked among livestock all his life conceded that he feels an emotional pang when he puts "*awa* [away]" an animal he is fond of: "Ah, well, that's *ma petty* [pet]. . . . *Ay* [yes], *ye* did *hae* a feeling, oh *ay*." Likewise, a stockman explained that he reacts differently depending on which animals are going away and the nature of his relations to the animals. For example, a bull might stand out because it had been on the farm for a number of years. Although "you'd never trust a bull," he said, "you probably would feel something the day [it went] away." But, he added, "If you let that bother you, you *cudna* [couldn't] do the job. . . . I have to load that animals onto the float, and as far as I'm concerned, they're on the float going somewhere. I know where they're going, but I *canna* let that bother me. As I said, I couldn't do the job. So you *canna* think about that; so you *hivna* got the same emotional attachment to them."

Hobby farmers are often new entrants to farming and have little experience slaughtering animals. Keeping small numbers of pet livestock can heighten the emotional intensity of this stage. For example, one hobby farmer described the emotional trauma of sending for slaughter two lambs that were so pet-like they followed her into the trailer. She regarded this as the worst thing she had ever done to sheep, as well as the most painful and regrettable: "I knew they had to be killed but it was the way that they trusted; I led them in to their death. I found that very difficult to live with. I still find it quite difficult; it hurts still to think of it, because . . . , really, I betrayed the sheep by leading them into the box."

Other hobbyists also dealt with feelings of betrayal and guilt. One woman, however, pointedly refused to feel remorse. Although it is difficult when a "real character has to be taken off to the slaughterhouse," she conceded, "everything's got to die in the end. . . . They wouldn't be here unless they were going to be slaughtered, and provided you give them as comfortable a life as possible when they are alive, then you've done your bit by them. I refuse to have any feelings of guilt about marching them off to the slaughterhouse. . . . [A] lot of people think it's very hard-hearted and worse, but you couldn't have fields full of geriatric animals."

As previously discussed, hobbyists virtuously assert that their animals have been "looked after properly" and have had a "good life." Perhaps such beliefs partly assuage any uncomfortable feelings and thoughts they might harbor about killing their stock. A few small-scale producers, however, were troubled about their part in truncating animals' lives. For example, a farmer's wife said:

> So long as the slaughter process is humane, the actual slaughter is not a welfare issue. What happens before that point is, but the actual slaughter, I think—once it's happened, that's that. So I don't think the animal really is suffering from that, providing it's had a valid life and it's had a reasonable time. . . . [I]f it wasn't going to be eaten, it wouldn't have had that. So I can justify it that way. But actually, . . . it is me causing this animal to have a shorter life than it would otherwise [have], and yeah, I feel guilty.

Interestingly, a hobby farmer whose child had died prematurely identified with this point. She explained, "I think it's so unfair that you get such a short spell in life, on earth, that [for] some people life is abbreviated. And I think the same about animals—that they've an even shorter spell and they have a happy little life, too. But I think it's unfortunate that we're cutting it short when they're at their prime of life."

Finally, some hobbyists eat their own animals, but not usually special ones. Some dispose of favored four-legged friends in a manner that is deemed more respectful and dignified than routine slaughter. For example, a married hobby farmer explained, "I quite like things to be functional; everything on the farm here that hasn't got a name is available for freezer goods. If it's named, we tend to give it a decent Christian burial." Similarly, another hobbyist said he had singled out his dairy goats and the eldest angora rabbit for "an honorable burial in the garden when the time comes" because he preferred "the dignity of a burial here. That's the difference between, if you like, my producing animals."

Generally, stockpeople and hobby farmers work with store and prime animals over a period of months and with breeding animals for a number of years, so they get to know some of them individually. This can contribute to feelings of unease. In contrast, mart and veterinary workers tend to interact with a fairly transient population of unfamiliar animals.[4] Despite this, such workers also can find the slaughtering phase problematic. For example, an auctioneer who procures lambs for slaughter admitted that he still hated to see the first spring lambs picked for slaughter each year go away. "They're so short *o* life. There's one farmer who does early lambing. They're born at the beginning of January, and then the first ones go away by the middle *o* April—twelve to fourteen weeks old. You've got to detach *yersel* [yourself]. . . . [I]f it's ready to go, *ye* mark it. If you're going to start worrying *aboot* it, you'd go off your head."

He also thought farmers selling animals through the sale ring might do so because "they *dinna* feel responsible that they've actually gone and killed that sheep or beast or whatever. They leave that *tae* somebody else to do; they're passing the buck on, more or less." The majority of farmers, he believed, "*dinna* like seeing their beast going away. They just financially got to do it. Most of them have got some compassion on the animal." A mart worker who processed livestock via the loading bays also cuts off from the slaughtering stage: "When I finish work at the end of the day, I don't think about what happens to them afterward. I can't, really." When I asked why, she said, "It's probably easier to think that it all stops here. They leave on the floats, and they all live happily ever after. Yeah, it's definitely easier to think that than to think that they're now away to do something else that doesn't involve living on a farm and eating grass." Because the worker constantly handled a transient resource of *live* animals, I wanted to clarify whether that mitigated her feelings about sending animals for slaughter from the mart. She said:

> Yeah, I think so. I think if they had a slaughterhouse next door and I actually saw them going in and heard what was going on, that would make life very difficult. It would make the job very hard. So yeah, . . . being with live animals is definitely important, 'cause then I miss the process, and I eat meat. So I sound a bit of a hypocrite by eating meat, but on the other hand, I want to support the Scottish meat industry, so I don't feel that I could become vegetarian, either. But I miss out that bit, the section where they're killed.

She was less perturbed, however, if an animal had to be shot at the mart because it was injured. "I find that far easier to cope with than if somebody shot it and said they were just away home to skin it and put it in their freezer." For the worker, casualty slaughter was more palatable than slaughtering healthy animals. She perceived injured animals to be in pain and believed that they tended "to know when their time's up." She had similar feelings about putting down pet animals: "That's the point when the animal probably thinks to itself, 'You know, get me out of here!' Whereas the killing for meat, the animal's live, healthy, I mean it's probably healthier than a lot, you know, that's the difficult bit." Although she appreciated that meat animals have to be in their prime to be killed, she mentally skipped over that point: "I suppose I just live in fairy world, really. I mean cattle come in here and [are] sold; then there's this huge gap, and all of a sudden meat appears on the counters." Thus, even though she worked in the marketing hub of the commercial livestock sector, she was able to partition off the animals she processed from the meat she consumed. Like those who are fully removed from producing animals, she adhered to this inconsistent belief. "I confine the whole process to what goes on in here and what happens afterward. I just don't or can't think about. . . . It just stops at the loading banks." Her deliberate strategy of not knowing, or her "discourse of ignorance," is made possible by the division of labor within the production system, which allows her to legitimately dissociate herself from the slaughtering stage by claiming, "It's not my job" (Mike Michael, quoted in Birke et al. 2007, 116).

Veterinarians, by contrast, are exposed to the killing of food animals. A newly qualified vet, for example, explained that she had attended the abattoir to observe the slaughtering process as part of the veterinary-school curriculum and that, since graduating, she sometimes had to go to the slaughterhouse as part of her veterinary duties. Based on such experiences, she said, "I don't like it [at] all. I find it just really a sort of production line, which it has to be, obviously. But . . . it doesn't [make] me squeamish; I just find it a very cold and gloomy atmosphere. People who work there I found very scary. . . . Well, you walk in and they all stare at you—you know, they're all holding knives. I just don't feel happy in an abattoir at all. It just makes me uncomfortable."

Part of her discomfort stemmed from seeing "lots of beasts hanging upside down, blood everywhere, and everyone's sloshing around with . . . bloody boots on, yielding knives, chopping things up. It's just quite a distressing place with all this death around." She was a meat eater and was not opposed to killing animals, as long as it was carried out humanely. Even so, she pointed out that supermarkets display meat products in neat packaging: "You don't see it in the shape of, you know, half a cow hanging on the wall." More to the point, seeing dead animals cut up into parts fundamentally challenged her image of animals: "I would rather see it [as] the final product of mince or steak or something without seeing in-between." For her, this was a "bit too close for comfort."

A senior veterinary student offered a more positive, contrasting account of visiting and working in an abattoir. Although she had dreaded her initial visit, she found the experience reassuring. Before the visit, she said, she had "never made the association between live animals and beef, and now I do. . . . But it doesn't bother me because I know how humanely everything's done in the majority of abattoirs in the country. As long as people want to eat beef, it has to happen, so it's important that it's done properly." She was impressed by what she saw, as were some of her fellow students. Even vegetarian colleagues who had to make the abattoir visit, she said, "have now started eating meat again" because "it wasn't anything like [they] thought it would be, and it's not cruel at all, and there's no reason not to eat meat." She also praised the slaughter workers:

> I actually quite admire a lot of them, because there's a lot of them who hate the job, you know, but they do it anyway, and they do it well, which I think has to be to their credit. If it's something you don't like but you still make sure you do it well, I think that's great. They are really well paid. I guess that's the reason a lot of them do it. You know, they're a great bunch of guys, really friendly. . . . And they still have the animal's interest at heart, as well—very impressive.

By claiming that most abattoirs in the United Kingdom are competent and humane, she implicitly acknowledged that slaughtering practices can be problematic.[5] Slaughtering animals is a highly emotive and controversial occupation, and her work required her to be part of this potentially suspect world. By insinuating that bad practice is perhaps more likely to occur in non-U.K. abattoirs, she morally aligned herself with "good" U.K. slaughter facilities while morally criticizing those employed in non-U.K. meat plants. Clearly, the demarcation of such clear-cut boundaries is overly simplistic. Even if all abattoirs were assessed and could be located on a continuum of animal-welfare and human-welfare standards, there would be great variation in how those standards were met among different facilities and within individual facilities. Nonetheless, the veterinary student's rhetorical strategy had created a "socioethical domain, outside of which are people who do not 'do things right' by animals" (Birke at al 2007, 116–117), but she perceived the abattoir she had visited as being situated within the domain.

Even so, having sung the slaughterers' praises, she was adamant that she could not do their job:

> I don't think I could cope with it. I cope fine with going out and putting dogs to sleep, cats to sleep, shooting cows—casualty slaughters—but I could not do that day in, day out. I think it would depress me too much, . . . I'm a vet, and it's my job to try to keep as many [animals] alive as I can, you know, and I couldn't ever . . . spend my whole day in an abattoir.

Producers shared the association between killing animals and feeling depressed. For example, a hobby farmer who worked in the commercial livestock sector asserted, "It's a job I couldn't do. I couldn't be in a slaughterhouse day after day. I find it actually quite depressing killing things." A former commercial sheep farmer (female) similarly explained:

> I actually think I'd probably end up feeling quite depressed with the thought that, day in and day out, my part in the production line was killing. I do think that eventually I would say, "Hey! wow!—have some birth for a while!" I don't think I could tolerate it. I might be able to do it for a week, but even then I don't think I would enjoy it. Maybe a lot of that's up here [*pointing to her head*]. I don't know. But I don't like the bit about being, you know—you're the one to see them at the very end, cutting them up.

Conversely, a commercial stockman who sometimes visited the slaughter-house to assess how well he had finished his animals explained that, once they had been turned into cuts of meat and were thus unrecognizable, he could "look at them in the hook, and that doesn't bother me at all." What he found objectionable was being able to recognize his animals hanging up at the beginning of the slaughtering process. The transition from livestock to meat involves an in-between stage in which the animal is technically dead but not yet meat. This state of limbo can be especially tricky for people to witness and negotiate. For example, a hobby farmer who reared and ate his own animals admitted that, when assisting the person who transformed his animals into something edible, "I would rather be somewhere else." However, he said, he found it "interesting in the sense that you start at nine o'clock in the morning with a live animal in the shed and you take it out, and by . . . eleven [or] twelve o'clock, you've got joints of meat, which are going into freezer bags." He also found the in-between stage awkward:

> Once they've . . . been shot and they're lying there and he cuts their throat, you have to hang them up, and they're a complete deadweight—literally. You hang them, and especially when he starts . . . cutting the head off and, you know, getting all the guts out and then skinning them and everything. Then all of a sudden it's just like a carcass in a butcher's shop. It's all right once it's like that. I'm much happier. I'm quite happy to cut it up, but . . . I wouldn't really want to kill it or skin it. But that's partly cause I'm a bit squeamish.

Those who consume their own animals know the animated source of meat on their plates. This can neutralize the distancing effects of dissociation strategies. The hobby farmer recounted a story about friends who had fielded questions from their son the first time the family ate one of their pigs for Sunday lunch. The boy, who was about six, asked, "Mum, which pig we eating today?" She answered, "Ophelia." They carried on eating a few more mouthfuls, then he asked, "Which leg we eating? Is it a front one or a back one?" She replied, "A back one, I think." They carried on eating. He then asked, "Is it the left leg or the right leg?" At that point, the mother gave up on her Sunday lunch. The hobby farmer and his wife identified with their friend's situation because their children had made similar comments. They handled such queries, they said, by encouraging the children "to accept that the animals are there only because somebody is going to eat them at the end of it. . . . So when they say, 'Oh we can't send them off to be eaten,' [we say], 'Well, you know, [we] wouldn't have had them in the first place. [Y]ou couldn't just have a field full of animals just looking nice, because something's got to happen to them at the end of it. I mean, that's why sheep are domesticated—because people want to eat them" (see Herzog and McGee 1983).

The Killing Job: Can You "Stomach It"?

Having described how agricultural and veterinary workers perceive and deal with the transition of livestock animals into meat, I now consider how slaughter workers perceive the process. As highlighted, byre-face workers often use euphemisms such as "going away" and "down the road" when sending their animals for slaughter. The meat industry also uses strategies of verbal concealment to "conjure up an image of meat divorced from the act of slaughter." Terms such as "butcher" and "slaughterhouse" increasingly have been marginalized as American euphemisms such as "meat plant" and "meat factory" have gained favor (Serpell 1996, 196). An assistant abattoir manager stated: "A killing house kills things." Even though the activities at a meat plant are the same as those at a killing house, people may be more comfortable with "meat plant," he said, because "they know exactly what's going on" in a killing house. "Meat plant" just makes the killing process less blatant. Moreover, he explained, he has changed the job title "slaughterman" to "meat designer," because "that's basically what they do." Informally, workers still use the term "slaughterman." Even so, the meat industry's participation in a linguistic makeover not only denotes its attempt to minimize offending public sensibilities but also seems to be part of an official strategy to assist in attracting, recruiting, and retaining staff.

Workers know that working in a slaughterhouse is not the best job in the world, but often they see it as an intermediate job until they get another position. For example, a mart worker said that he had worked in a slaughterhouse for two years after he lost his job. A married man who wanted to leave offshore work was employed at a slaughterhouse for four months until he got another job. The assistant manager said that slaughterhouse work was an employment option for men who do not have an education—those who, historically, often had worked in "the railway yard, the building site, and the meat trade." He noted that one reason the job was done by men was that, until thirty years ago, all the slaughterman did was kill the animals. "Then you had laborers who dragged it out [to] the road, chopped it up, whatever," he said. "It's very heavy work, then, really heavy work. More strength than brains was needed. I think the males of the species would have made it very macho at that time."[6] Furthermore, slaughtering is a "low status, routine, and demeaning job" (Thompson 1983, 215). It is noisy, smelly, dirty work, and workers can be stigmatized for what they do: "Work is one of the things by which he is judged, and certainly one of the more significant things by which he judges himself" (Everett Hughes, quoted in Arluke 2006, 21). One worker described the sensory nature of this work context: "It is loud. . . . *Ye've* got the noise of the box—the minute a beast goes down, *ye* have *yer* clatter. Ye have the shackles coming down the slide; it's noisy. *Ye've* the big *saa* [saw], *ye* hear it. After a while *ye dinna* hear *naething* [nothing]. *Ye* get accustomed to the noise. . . . There's the smell of the blood, . . . hot fat, *ken* when a beast is shot it's still hot, *ye* feel the heat come *aff* the actual beast."

When asked how people might perceive slaughter workers, he said, "Well likes *o yer* animal rights think of us, us murderers. But if they were ever to go in and see it done, it is not a cruelty, it's all humanely done. . . . *Naebody* [nobody] says, he *wis* a dirty rat or *onything* [anything] like that. [But some people] do back stab: 'Oh, yes, that's him that kills *aa* [all] the beasts.' And the folk that's *deein* that is the animal rights and *yer* vegetarians." He also noted that sometimes people said, "'I'm *nae spikin tae* you. Why? Oh, you kill cattle. Ye kill sheep.' I say, 'Well, you like them. How you going *tae* get them if . . . there's *nae* kill?'" Another worker said he had had lively discussions with vegetarians, while other folk saw slaughter work as "just another job." Some balk at the idea of "blood and guts," he noted, but perhaps they would react similarly to a nurse in an emergency ward. However, he also said that, if he knew a person was squeamish, he would brazenly announce that he worked in a slaughterhouse just to get a reaction. The assistant manager made a slightly different point: "If they're a bitty uppity or anything else, then that's a target for me to knock them down."

As noted above, retaining staff can be a major issue in this sector. The assistant manager of a specialist abattoir employed eighty-five slaughter staff who aimed to kill and process 1,000 cattle and 800 pigs per week. Until a few years ago, the plant was failing to retain workers, especially good ones.[7] "Up *tae* then, we didn't do a lot," the assistant manager said.

"You arrived here in the Monday; you got your boiler suit, apron, and pair *o* [boots] and a hat and [were] shown the job for about five minutes, and that was you—sink or swim—which happened for years." The abattoir's management reviewed its recruitment and retention practices and has since introduced a number of changes, including a two-day induction for new recruits to orient them to the plant and the allocation of a "minder" who is responsible for integrating the new hire into the work environment. The minder also deals with questions about his job, health and safety, and so forth. New laborers always start by sweeping the floors and tidying up. If they show any interest, they will then get a job on the slaughter line:

> Very first job is washing insides. Each job after that is a bit more skilled, and it also means they get a . . . larger share of the bonus. The more jobs you can do, or skill that you have, the more money you get. The guys who are 100 per cent, the top guys, can do all the other jobs. . . . They know the problems; they know if it's a good [or] bad . . . job. That's where they earn their respect from the other guys on the line.

Job rotation has been introduced to deal with boredom, and the speed of the line has been slowed down: "If your line speed is too fast, the guys have nothing else to do but get their head down, get the job done, and it's a job. It's not a way of life. They can't crack a joke between them or whatever. . . . [I]t's mind-bending; it's boring. It's, 'I'm not going to be here very long; I'm going to look for something else. I'm not interested in the actual product. It's just a job.'" Because the specialist abattoir's target market is high-class butcher shops and restaurants, the assistant manager explained, "We pay a lot of money for the animals. We expect a lot of money for the product. We expect quality work. The only way you get that is quality staff. How do you get quality staff? You've got to train them. Work on human nature, as well, a *wee* bit."

However, not all abattoirs target the upper end of the market or are concerned about creating a work ethos that minimizes boredom to retain staff, as a former slaughterer noted. He had worked on more manual and mass automated slaughter lines. He preferred the manual line because it was slower and because he was not restricted to "one little job"; having responsibility for "two or three different things" introduced some variety into his working day. When he started, he noted, the abattoir had no automated line for lambs: "*Ye* just like killed one at a time. You set [a lamb] on this table, basically, and skinned it, manually lifted it to each station. And then they got a motorized line whereby instead of the slaughtermen moving around the carcass, the carcass moves past the slaughterman; he just [stood] in one place." He described the slaughtering process in more detail:

> The first thing that's done with the lambs, *ye* electrocute them with tongs, *jist* like scissory things that you pick sausages out of a frying pan with. Big things like that with a[n] electric cable attached to it. Stun them *wi* that. And then they get winched up—it's an overhead line—and the slaughterman slits their throats. The next stage, there's the hide around the legs is likely to be slackened off. . . . [Y]ou get a slit up the front to open the skin back, and then it's hooked up to a machine that pulls the fleece off in one go. And somebody slits it up the middle, pulls the guts out—the organs like the liver, the heart, and that. The head's cut off.

In contrast, working on the mass automated slaughter line was so boring that he left after three months. He described the line as "massive": "There's just this big chain from the bottom end right up to top and then back around and down again past the chills on the other side of this wall.[8] There was about thirty of us on that slaughter line. [We] just stood on that line side by side— enough room for your elbows, basically—and the lambs just went passed you. You had ten to fifteen seconds to do your little job. You just had eight hours a day. It was . . . so boring that you switch off from the job. [I was] left with so much time to think about, Why am I standing here doing this? It's the futility of it." Eventually, he "jacked it in," even though that left him unemployed.

Another slaughterer who worked in the "kill" described it as a factory "where animals came in live and they got *killt* [killed]." He had received no specialist training before doing what he called the "basic job":

> It's like you're in a factory. . . . Can you imagine [being] in a room and the pigs are, like, going around like [on] a train, but [they're] hanging upside down, and it just constantly moves all the time? And there's hooks. The pigs are suspended, like, ten feet up in the air, and there's platforms for the guys to work on. Once they cut open the pigs, they take out the heart and the liver and the lungs and the intestines and everything, and they hang the heart and the liver up, and they put the intestines in a basket kind of thing. My job . . . on my very first day was to take the hearts and the livers off of these hooks and hang them up on more hooks so they could be transported. Once this thing was full, they could transport them into the chill to keep longer. I can always remember the guts; the intestines went past me, and I'm going to be sick *wi* this, just *wi* the smell.

He likened the smell to decaying grass or dung: "My granddad used to like save up *aa* the rubbish and *aathing* [everything] in one corner of his garden and then use it as compost. It was that kind a smell. It was a really strong, really pungent smell." He also remembered "holding the lung, *ken* it was squingy, it was a spongy." It was common for new starts to do such jobs, he said, "just to see . . . if they're going to last in the kill, because if you were a bit squeamish or something like that, then the day to get *oot* is your first day."

These accounts illustrate how less mechanized slaughter lines incorporate workers into more stages of the process than mass automated lines, which depend on a highly fragmented division of labor that de-skills the workforce. One effect of working on an automated line is that the worker's "reality is apprehended in terms of components which can be assembled in different ways, there is no necessary relationship between a particular sequence of componential actions and the ultimate end of these actions" (Berger et al. 1973, 27–28). Hence, they are less aware of the animal and what is happening to it, apart from their specific input into the overall process. On the more manual line, the first former slaughterer said, "Everybody was involved with the whole process, so I suppose [we] were probably more aware of the transition that this carcass was going through." Of course, a slaughter line does not assemble a product. It disassembles and dismantles an animal. "The slaughterman dresses *une carcasse,* not an animal, because as we have seen the animal disappears in the act of suspension. As used here, the term 'dressing' never ceases to surprise, since the process referred to bears more relation to undressing" (Vialles 1994, 49).

Legally, before any animal can be bled and disassembled in the United Kingdom, it must be stunned (Gregory 1989–1990).[9] Historically, one of the reasons for instituting this measure was not to minimize the animal's pain, but to ensure that pre-slaughter animals could not view other animals undergoing slaughter (Gregory 1984; King George V 1933, 9). The timing of the earliest legislation coincides with a number of debates in the 1920s and 1930s that deliberated traditional methods of slaughter and the desire to promote more humane slaughtering practices (Walker 1986, 127). For example, G. Wooldridge (1922, 139), a veterinarian in London, regarded this "as one of the burning questions of the day" and advocated that "slaughter should be carried out in such a manner as to prevent all avoidable pain and suffering, both before and at the time of slaughter." He believed that the veterinary profession was best qualified to assess this aspect and thus contributed to legitimating its expert status and input in this emerging area. Vets have become an integral part of the food-animal chain. Joanna Swabe (1999, 154) writes that the vet is "one of the most important cogs in the wheels of the complex machinery that is responsible for the transformation of living flesh into animal produce fit for human consumption" (see also Huey 1998).[10] For example, veterinarians and Official Veterinary Surgeons (OVSs) monitor and manage livestock diseases to ensure disease-free meat and are responsible for examining animals before they are slaughtered and meat after they are slaughtered.[11] A veterinary student described what an ante mortem examination might involve:

> It includes things like how clean the animals are—that's important for your hygiene, especially after the E. coli outbreak—and just that they are showing no physical signs of illness. They don't have a high temperature, and [they're] not lame—not just that they're not lame; they have to be weight-bearing on all their legs and [have] no obvious swellings and things like that. So it's the vet's duty first thing in the morning, say half past six, [to go] into the lairage, and they have to individually check each pen and check each animal. Anything that looks slightly off has to be put into the crush and thoroughly examined—heartbeat, respiratory, temperature—just to filter out the ones that aren't fit to be there and therefore probably aren't fit for human consumption. Then, of course, anything that maybe is borderline, a bit lame but not too bad: They organize their slaughter rota so that they'll go first, so they're not delaying slaughter unnecessarily and they're not going to be suffering any longer than they need to be.

Assessing the cleanliness of pre-slaughter animals is a fundamental meat-hygiene issue. Since the Meat Hygiene Service was formed in 1995, it has played a pivotal role in protecting "public health and animal welfare through the fair, consistent and effective enforcement of meat hygiene, inspection, Specified Bovine Offal (SBO) and animal welfare legislation" (Soul 1998, 64). Following an outbreak of E. coli in England a year later, the Meat Hygiene Service established the Clean Livestock Policy (CLP). An OVS explained:

> The whole reason for the CLP is that when the animals are hung up in the abattoir, the first point of incision is often where the E. coli can get onto the carcass . . . and it tends to be at the entry points when one is skinning an animal; it tends to be on the knees and on the hocks, where often the dirt is adhered to, and also on the brisket and along the belly. . . . It maybe doesn't matter too much if there's a bit of sharn [cow dung] along the back of the animal, but where it's on the knees and along the brisket and along the belly, that's crucial. These parts of the carcass tend to be where the mince is created from; it tends to be the lesser-value cuts of meat. So that tends to be the product that goes into mince and tends to get into burgers and so on.

The CLP is illustrated in a poster that displays five grades or levels of livestock cleanliness, whereby cattle graded five would be covered with *sharn* from head to toe; grade four would be pretty bad, too; grade three is borderline; and grades two and one are ideal levels. Borderline animals might be kept in the slaughterhouse lairage overnight; during transportation, some animals can become quite messy, but if they lie on straw, they dry out and by morning may be clean enough to be slaughtered. Grade four and grade five animals are sometimes returned to the farmer to be cleaned up. As Noelie Vialles (1994, 35) notes in her study of French abattoirs, "The 'dirty sector' is the realm of the warm, the moist, the living, of smells and secretions, of the biological threat that needs constantly to be contained and cleaned up. The 'clean sector,' on the other hand, is where everything is inert, bloodless, trimmed, and stabilised by cold." Between the dirty and clean arenas the animal is stunned and bled before it is dismantled. Since some workers cannot "stomach" the initial stages of the slaughtering process, it has possibly become an informal way to assess new entrants to see if they can handle it. For example, a mart worker who once worked in a slaughterhouse described his first day killing cattle: "The head linesman came *tae* me in the morning and took me up on *tae* the box, and said, 'Shoot the first beast.' It *wis tae* see if I could stomach it. *Dinna* throw up." He continued:

> If *yer* getting a beast coming in *tae* a box and the beast is actually looking at you straight in the face, you're *stannin* there *wi* a gun in your hand, and it's thinking, "That's a dirty sod, that." The boy says, "Can *ye sheet* [shoot] it?" *Ay*, bang! He says, "How do you feel?" I said, "*Aa* right." So he took us *doon*, and once you shoot them, you've *tae* spring them, which means *pitting* [putting] a metal spring into the hole in the head, which knocks *oot* the *hail* [whole] spinal column, it knocks *oot* the nerves. And once that's done, you've *tae* shackle them, which means *pitting* a chain on *tae* their hind leg and whip it up on *tae* the rail. *Dee* [do] that, *tak* [take] them *roon*, cut their throats, and bleed them. He said, "How you feeling?" I said, "I'm fine." He says, "OK, carry on, *dee* the next thirty and we'll see how *ye* feel *aifter* [after] that."[12]

A man who cannot do it might get a job farther down the line: "*Ay*, maybe *pit* them skinning, or *doon* the far end, actually once the beast is *splut* [split]. Maybe taking the liver and *aathing oot, faar* [where] they're *nae* seeing the actual beast being shot." Similarly, a former slaughterer suggested that he had been "egged on" by colleagues to shoot a cow, even though this was not part of his job: "I was a sort of new boy there, or something like that, and I did it. . . . I felt a bit sort of eeky [squeamish] about doing it, but it wasn't that big a deal." Another job he saw new people being asked to do was "*slitin'* [slitting] the throats." A former worker described seeing pigs suspended on a line above a bloodbath getting their throats cut: "Initially when you cut them, there is a big flow *o* blood comes out, and they would . . . basically let the blood out." The bloodbath amazed him, he said, because "it was a huge area. . . . By the end of the day, it was so full of blood, I *jist dinna ken* there was *sae* much blood." The blood was pumped into four or five forty-gallon barrels, and although he was unsure what happened to it, a company collected the barrels daily. Clearly, the bleeding process signifies the loss of vitality as it drains from the animal's body and ushers in its impending death: "To bleed an animal is indeed to take away life itself, the vital principle, therefore to de-animate" (Vialles 1994, 73). However, as one of the workers noted, if a person failed or declined to stun or stick an animal, he ran the risk of being branded a "woose," or somebody who was not as tough as he appeared to be. He described the workforce as a fairly tight-knit group and said that such challenges were indicative of "bonding" or "barrier breaking." Similarly, new recruits to animal-shelter work might experience "moral stress" after their initial exposure to euthanasia (Arluke 2006, 116). To some extent, "Learning to kill animals was a rite of passage that marked the transition of shelter workers out of the role of novice" (Arluke 1991, 1178).[13] This seems applicable to the abattoir setting, too.

Quite simply, a former worker said, "Some folk *jist cudna dee* it. It was *jist* a case a blanking yourself off and getting on with *fit yer deein*. Because if you *wis* to sit and think *aboot* it, and let it play on your mind, I suppose you *wid* stop *deein* it." As the assistant manager explained, "Nobody likes to do it [slaughter animals]. So they've got to, they don't have an excuse for doing it; it's a job." Of course, he noted, people ask, "'How could you do that?' The only reasons that they have for it—the only reason that they actually do it—is it is a job. But to justify it a *wee bitty* and show that they're not too emotional, they would try and make it butch and macho." When asked whether any part of the process was most likely to induce emotional responses from workers, he replied, "Shooting the animal. I can do it. I don't like doing it. A lot *o* guys are the same. A lot *o* guys can't do it because you have to face the animal and shoot it at point-blank range." The significance of being face to face with the animal was captured in a "moon-eyes" look, which, he explained, was a "sad, longing, unsure, looking-for-comfort [look]. You see that walking towards you and you've got to shoot it." Before it is shot, the animal is a living thing, he said. But once the animal had been shot, it became a "unit" for him, "like a tin *o* beans. It's *jist* something I've got to do work on now. There's no emotion attached to that. It's a thing now. It's not an animal." This process of de-animalization occurs as soon as the animal's feet no longer touch the ground: "Between the moment of death and the final presentation of the carcass, there is a nameless void: the object is neither an animal, not even (especially not even) a dead animal (a corpse, unfit for consumption), nor is it yet meat" (Vialles 1994, 44).

Given the division of labor underpinning the industrialized killing of animals, Vialles (1994, 45) asks, "Who kills the animal? The person who stuns it, or the person who bleeds it?" She continues:

> Not only is such a doubt formally possible; it exists in reality. When asked, some will say that the bleeding alone causes death, which is true, but they will promptly add that, once stunned, the animal feels nothing; "it's as if dead," and bleeding merely finishes off a death that would in any case not be long in coming. Others consider the stunning crucial, and the reason they give is *the same one:* "it's as if dead . . . and what follows can no longer matter to it."

By splitting this role, the killing act is made nebulous. As Vialles suggests, this may assuage any traces of residual guilt experienced by those involved. Then again, the division of labor underpinning the production-slaughter-consumption process may be an inadvertent exercise in shifting the blame: "In a sense, everyone involved is guilty, but no one is obliged to shoulder the full burden of responsibility" (Serpell 1996, 204). Such ambiguity is possible in modern industrialized societies because livestock and slaughterhouses are generally out of sight and out of mind; having been banished to the margins of urbanized society, they joined the ranks of other socially unpalatable institutions such as asylums and graveyards (Philo 1995). However, perhaps we should not assume that being in sight, and having insight, necessarily means in mind.

"Pitting It oot o Its Misery"

Generally, the division of labor structurally differentiates the roles of producers from those of slaughterers, and thus livestock from deadstock. Obvious exceptions are commercial and recreational producers who produce, kill, and eat their own livestock. Even though the majority of livestock will be institutionally killed, some animals die outside the slaughterhouse too—for instance, as a result of casualty slaughter or livestock culls. Less routinized killing not only provides some insight into how producers who normally identify themselves with rearing and working with live animals respond to death in their own farmyards, but it also shows how vets, who perform euthanasia, perceive and manage putting livestock out of their misery. For example, an experienced hobby farmer commented that she was "always trying to save something's life. Although I could kill something . . . because it's so sick, I would have no qualms about something that was suffering. . . . But I'm not very good at killing something that's perfect[ly] healthy." This is a fairly common sentiment expressed by commercial and recreational producers. Nonetheless, putting an animal out of its misery also can be a distressing experience. A mart worker who had once farmed on a small scale explained how she felt about killing a lamb that had been attacked by a crow: "It had its eyes removed, its tongue removed, and there was nobody in sight who could give me hand. So I had to bash it over the head with a rock to put it out of its misery, and I just cried. I found it so hard, even though it was lying in agony, [and] it would be better off dead. There was nothing that could have been done. I just felt it was a waste. I felt I was a murderer, um, really hard."

Playing God, or deciding when an animal's time is up, is no easy decision, especially if the animal is old. The hobby farmer quoted earlier who did not feel any guilt about putting her animals to slaughter said she was equally comfortable putting an end to an animal's suffering. However, "Once or twice I have thought, oh, I did that too early if it was an elderly animal," she said. "But then you just have to say, 'Don't be ridiculous. It had to die sooner or later.' And you don't know. You never, never get it right. You'd never, ever do it at the right time; you're either too late or you're too early." Similarly, a commercial farmer said that, when looking after sick animals, she would reach a stage where she was "fighting a losing battle, and the kindest thing, because you have to be cruel to be kind, [was] to shoot it and put it down." Once the animal was dead, she could dispose of it, but she did not want to "be there when it's looking at me, when the vet's there to shoot it. . . . I'm a female, and I'm the one that has been closest to it this last while when it's been ill and I've failed it. . . . I haven't managed to make it recover from whatever illness [was] ailing it." Old cows can be especially problematic because a decision has to be made as to whether they will be bulled again to get them in calf. A cow that is past its level of productivity can develop complications that would have been avoided, she said, if it had just been "put . . . down the road when she could walk on a float and she could generate the magic £300. Instead here's a lot of work in front of me, and it's my own selfish fault." However, she noted, it was her husband who tended to push such animals to their limit:

I'm always saying to [my husband], "We're not bulling that cow again; she's come to the end of her life." And he always says, "*Ach,* she'll do another year." So it's always me that has to look after the older ones and nurse them along. And it's always me that has to make the decision: She's off her feet; she's going downhill; I'll have to give up on her. And I could have made her life easy by not putting her in calf, because that's a strain, and letting her walk on the float and just being killed in a knackery somewhere.

This example not only reveals a fairly stereotypical gendered division of labor. It also raises two additional points I explore further. First, vets kill casualty livestock on the farm and "put down" companion animals. However, the management of these two types of domestic animal death is very different. Second, producers can experience a profound sense of personal failure and anguish when their animals do not recover or die on the farm.[14] I now provide examples of each aspect.

Killing animals is an integral and routine part of veterinary practice. Euthanasia is "one of the most important things that we as vets are able to do," a veterinary student said, "because we've got the power to take away the pain and suffering." In this context, vets are responsible for "inducing a pain-less death" (Jerrold Tannenbaum, quoted in Swabe 2000, 301). But, as Swabe, notes the term "euthanasia" usually applies only to the "deliberate killing" of pets and horses. Livestock animals are slaughtered. How clear-cut is this semantic distinction in practice? The euthanasia of pets is characterized by the vet's desire to make the animal's death as painless as possible while mini-mizing the level of emotional distress for the owner or guardian. This is of less concern when it comes to farmers and their animals. As a vet explained:

If you're injecting [a pet animal], . . . you get a vein and you get the stuff in, in a oner. . . . [Y]ou don't have a situation where you've got half of it in and the animal starts to struggle or it's fighting against you, and you're getting distressed, the owner's getting distressed, the beast is distressed. It just turns into a horrible situation. When you're shooting [a livestock animal], it's just a case of shooting it and not getting kicked when it falls down. But definitely, with the small animals, with the horses, it's more [about making] it as easy for the animal and the owner as possible: nice and smooth and non-distressing.

Her rationale for carefully managing the death of pets, as opposed to that of livestock, was that the owners of small animals and horses "are so much more emotionally attached to their . . . animal[s]." If they perceive "their animal is suffering or distressed, then that obviously distresses them an awful lot. You're in the position that you're trying to minimize everyone's upset." A vet-erinary colleague reinforced the importance of ensuring that the euthanizing of a pet is as "stress-free as possible." Killing livestock on the farm, however, is very different; it is much less emotional. It's more [about] money. Especially with BSE, . . . all we seem to do just now is write certificates for cattle that are over thirty months old to get slaughtered, either because they're not in calf or because they're just getting old and not doing so well. . . . Basically, if the farmer's cow is worth nothing, it's not worth spending any money on it. So if there's a chance it might not get better, [the farmer would] rather just send it away and get some money, . . . [w]hereas if it died on the farm, [the farmer would] get nothing.

She noted that her preference would be to investigate the animal's condition further, but the farmer might not want to do that because it would cost money. "He's not going to get the money back, and he's got no guarantees. So that is very frustrating." Negotiating the care of a "virtual patient" is challenging in both small and large animal practices because it requires the animal's owner and the vet to collaborate to establish the nature of the problem and how best to proceed. Differences of opinion can and do arise when lay and professional appraisals of the situation do not match up. As Clinton Sanders (1994) notes, vets often decide whether a client is good or bad based on how he or she balances or prioritizes the economic cost of treatment in relation to an animal's welfare. In this case, the vet perceived the farmer's decision negatively. But at the end of the day, the vet can only advise.

In terms of shooting the animal, she explained:

I suppose it's just something that you do. You get used to it. I didn't like it the first few times I did it. It's quite, uh, you know—I hope the animal doesn't suffer any more than it has to, because things can go wrong. You might have to shoot them a couple of times. It's not nice when that happens. Ultimately, it's the welfare of the animal I'm thinking of. If there's a cow that can't get up, it's been down after calving and it's not going to get up, then the best thing for it is to be euthanized. But when you have to do healthy cattle . . . you decide that this shouldn't be happening. You know, there's something wrong with the system somewhere that you have to do this, and that annoys you a little *bitty*.

The vet perceived killing casualty livestock as a form of euthanasia, as opposed to slaughter. Perhaps this is most likely to occur when a productive animal has transcended its tool-like status. For example, I accompanied another vet on a farm visit to find a cow in the midst of a difficult calving. On arrival, we found the cow lying on her side between a gate and the barn door. The vet tried to calve her normally but decided a caesarean section was required. When the calf was delivered, attempts were made to help it breathe. Unfortunately, it was too late. The cow had been in labor for some time. The farmer lost the newly born calf and would lose the cow too. On that occasion, the vet killed the cow by injection. The farmer had insisted that the cow's death be made as painless as possible, because he was particularly fond of her. When we left the farm, the vet remarked how upsetting the farmer had found this situation. His response had affected her, too.[15]

Some of my findings resonate with the experience of animal-shelter workers (Arluke 1994). For example, both sets of workers perfect their "technique" for killing animals because this diverts attention away from "why it needed to be done or how they felt about doing it—workers could reassure themselves that they were making death quick and painless for the animals" (Arluke 1994, 151). The alleviation of suffering of injured or old animals was also more palatable than killing young healthy animals. The economic fallout of BSE led to the introduction in 1996 of the Calf Processing Aid Scheme, which basically paid dairy and beef farmers to "cull newly-born calves in an effort to reduce cattle numbers" (Press and Journal 1998, 23).[16] The "Herod scheme," as it was also known, incinerated male calves usually exported for veal production to Europe.[17] When the export market folded, it left a glut of unwanted and unmarketable calves in the United Kingdom. The mart became a designated collection point and was responsible for coordinating, screening, and processing eligible calves for incineration. Only calves seven to twenty-one days old and deemed to be in good general health could be incinerated. For example, a worker explained, it was important to check the newly born calves' navels for infection, wetness, or any other abnormalities. The mart did not accept calves with infections because "the government are *nae gaan* to pay *oot* money for calves that are *nae* in good health." Even though the worker was "near in tears" while loading calves onto the float, she knew that it had to happen: "I mean there is *naething* [th]at me or you or *onybody* could *dee tae* stop it. So as far as I can see, the best I can *dee* is *mak* sure they're looked *aifter* properly. But I prefer *nae* to think *aboot* the calves *gaan awa tae* be killed an that. I mean, I *jist dee ma* job and *mak* sure they're being looked *aifter*."

Adopting a caretaking role is one way to deal with the stress of situations such as this one, and this worker believed she was the best person for the job because of her diligence in attending to the animals (Arluke 1991, 1179). But sometimes she made sure she was not around when the time came to load the calves, because she felt she was killing them by putting them on a truck. This vividly illustrates what has been called the "caring–killing paradox" (Arluke and Sanders 1996, 85). Another worker explained, "My heart bleeds for [the calves]; it really does. . . . It really is a waste of life. . . . They're looking for milk, and they're trying to suck your fingers, and what have you. They're ever so sweet, and I know they grow into monsters, a lot of them. But . . . they're just babies, and why breed? . . . I find that really difficult to cope with: you know, youngsters being killed like that."

These animals are neither ill nor injured; they are casualties of a malfunctioning system. In some ways, the death of healthy young productive animals is a routine occurrence. Meat animals are slaughtered every day in the prime of their lives. Are workers simply grappling with the economic wastefulness associated with discarding potentially productive animals? Or is there more to their troubled responses? Media images of emotionally distraught farmers dominated public consciousness in 2001 as the U.K. government tried to manage a devastating outbreak of foot-and-mouth disease.[18] Farmers not only observed firsthand the systematic culling of entire herds and flocks in their own backyards; they also watched their deadstock smolder on the numerous pyres dotted throughout the countryside, especially in Cumbria (Convery et al. 2005). The "contiguous cull" policy contributed to the slaughter of more than 10 million "mostly healthy animals"; this contentiously legitimated the killing of "all susceptible animals within a 3 km radius of a confirmed case" (Woods 2004, 342). However, the media images also bewildered some commentators: "It was tempting to see the many farmers who openly wept at the slaughter of their stock as simply hypocritical, since these animals would inevitably have been sent to identical deaths within weeks or months" (Smith 2002, 53).[19] Slaughter is indeed part and parcel of livestock farming, but "few farmers and livestock handlers would volunteer to be at the interface between live animal and food" (Convery et al. 2005, 105). Ian Convery and his colleagues (2005, 104) provide one way to make sense of the emotional, social, and economic shockwaves that reverberated throughout and beyond the farming communities caught up in the seven-and-a-half-month epidemic: "Death was in the wrong place (the farm rather than the abattoir), but it was also at the wrong time (in relation to the farm calendar) and on the wrong scale (such large scale slaughter seldom occurs at the same time)." In terms of my study, death may simply have been too blatant for those used to handling live animals, especially since it involved cute, neotenized calves that can engender "an adult nurturing response to such a 'loveable' object, eliciting feelings of tenderness" (Lawrence 1989, 62; see also Serpell 2003). Perhaps these workers could not fully support the government and bureaucratic rationales used to justify what they perceived as the senseless killing of otherwise healthy young animals. Even though producing and killing animals for meat is far from straightforward, many employed in the industry have learned to accept this rather unpalatable fact of agricultural life from a relatively young age (Ellis and Irvine 2008). What the cull policies associated with foot-and-mouth disease and BSE have shown, albeit for different reasons, is that killing livestock for non-edible reasons can be problematic for byre-face workers. The extent to which this reveals the fragile nature of taken-for-granted mainstream agricultural norms remains to be seen.

In sum, producers and slaughterers need each other. But their occupational interdependence does not necessarily mean that producers value the work of their killing colleagues. What my data show is that for many byre-face workers, handling and managing live animals is an integral part of their professional status and moral identity. It can also be an important "distancing device" (Serpell 1996, 187). Because they constantly rear, market, treat, or finish an ongoing supply of livestock, this can contribute to their ability to deal with the inevitability of sending animals to slaughter. Usually, when a truckload of animals leaves, another cohort arrives. This partly explains why the extensive culling during the foot-and-mouth disease epidemic was so traumatic for some farmers: They not only witnessed *all* of their animals being killed on their premises; they also observed that no replacement animals filled the void. On that occasion, there were no live animals; there was only deadstock. Routine slaughter is traumatic enough for some, and most avoid thinking too deeply about it all. Moreover, the discourse of caring for and looking after animals properly enables commercial and non-commercial workers to extricate themselves, albeit to varying degrees, from the end stage of production. By doing this, they carve out socio-ethical sub-domains within and outside the commercial sector whereby they can hold on to the belief that they have the animals' best interests at heart, despite sending the animals to slaughter. In other words, killing is not their job. Farmers are producers. As a senior vet put it, "Their purpose in life is to try and keep the maximum number of their livestock alive."

However, just because abattoir staff are employed to kill animals, this does not necessarily mean they are indifferent to killing. Although slaughter workers may perceive their most negative critics to be located outside the production process (i.e., vegetarians and animal-rights groups), there is some evidence to suggest that they have critics within the system, too. This clearly begins to undermine any polarized debates between "them" and "us," since those working at the byre face and the slaughter face also grapple with the realities of killing animals. What unites all workers is that many struggle with, and do not have the stomach for, the actual point of transition from life to death to meat. The in-between stage has the potential to cause the most distress and ambivalence. The notion of humane slaughter is an important thread running through many of the accounts in this chapter. Increasing regulation of the slaughtering process has been motivated primarily by a concern to limit or remove unnecessary cruelty from the system and reflects a genuine, albeit contested, concern for improvements in the necessary end-point of production. Perhaps the idea of humane slaughter is important to allow those who work with livestock and deadstock to reconcile themselves to this endpoint. Of course, animals are not inanimate objects that can be routinely processed (Smith 2000). They are sentient beings that react to the situations to which they are exposed. Ideally, humane slaughter is the template for acceptable slaughter, but in practice it is more difficult to ensure. Livestock production is built on economics and an affinity for working with livestock. This fundamental tension means that commercial farming and hobby farming are ridden with conceptual, emotional, ethical, and practical contradictions. Those working in both spheres have to learn to negotiate a pragmatic and precarious path through the relatively ambiguous territory of human–livestock relations.

9

Taking Stock

*Food Animals, Ambiguous Relations,
and Productive Contexts*

> A peasant becomes fond of his pig and is glad to salt away its pork. What is significant, and is so difficult for the urban stranger to understand, is that the two statements in that sentence are connected by an *and* and not by a *but.* (Berger 1980, 5)

Most people in modern industrialized societies choose to eat animal-derived protein. Thus, for many, but not all, consuming meat is a legitimate cultural norm. However, the task of killing food animals is generally regarded as "'dirty work' . . . an undesirable and repugnant job" (Thompson 1983, 215). Clearly, the transformation of animals via slaughter into edible flesh is an inevitable and pivotal part of producing meat. In practice, this stark productive fact makes sense. Morally it can be less straightforward.[1] For example, many consumers avoid thinking about this in-between stage because it triggers varying degrees of cognitive and emotional unease. But those within the commercial and hobby livestock sectors can grapple with this stage, too. One farming contact said that slaughterers are little more than "animals," which implied that producers are more human. That farmers can distance themselves from the killing stage provides a practical basis for making such a claim. Drawing such an explicit boundary between animal production and slaughter enables non-slaughter workers to carve out a socio-ethical domain, or "moral haven," that effectively insulates them from a more problematic part of the process (Birke et al. 2007, 158). It also indicates how slaughterhouse workers can be a convenient target for other workers' discomfiture. Clearly, the division of labor reinforces this distancing strategy, because pre-slaughter workers identify with live animals, not dead ones. In other words, the essential but deviant slaughtering role runs the most risk of acquiring a "spoiled identity." Thus, those who carry out this role are most likely to be perceived by both farming and non-farming "normals" as less than fully human (Goffman 1963).

It is advantageous for us to assume that "bad" people do bad jobs, or that "stigmatized occupations attract certain kinds of individuals, who because of their psychological or social characteristics contribute to the occupation's reputation" (Davis 1984, 234). Slaughter workers are mostly aware of the general contempt that surrounds their job because they, too, are socialized into the dominant cultural values. Public debates about the welfare, rights, and sentient status of farm animals are also gathering pace. This moral quickening is introducing a critical edge to the appraisal of once taken-for-granted animal-farming contexts and practices. Those responsible for overseeing food animals when they are alive are not only conscious of this moral shift; they are also subject to increasing levels of personal and professional scrutiny. As Lynda Birke and her colleagues (2007, 154) note, "Some forms of work are now stigmatized that once had merited greater prestige. People in these occupations might now find themselves the target of moral crusades by groups seeking to change public opinion about whatever they find offensive. . . . Lurking behind such moral criticism are often implicit charges that these workers must be unprincipled or shameless to do what they do." Thus, byre-face workers, who previously enjoyed a fairly positive reputation, are now portrayed as being engaged in morally dubious work.

The move away from family-run farms and the problems associated with intensive methods of animal production undoubtedly have contributed to this negative trend in public opinion. However, producers work in a variety of productive contexts, with different species and breeds of livestock. As some approaches and stages of production are regarded more favorably than others, byre-face workers have an opportunity to reclaim some personal and professional integrity. In other words, some producers are differentiating themselves from others, whom they perceive to be associated with or tainted by more disputed parts of the process (see Holloway 2002). Self-appointed "good" producers are creating socio-ethical sub-domains to distance themselves from comparatively "bad" productive contexts and practices. For instance, my hobby farmers were keen to dissociate from the methods and attitudes of commercial farmers. By claiming to be more traditional, natural, and welfare-friendly than their farming counterparts, they believed that their animals, unlike those in the commercial sector, enjoyed a "good life" (see Holloway 2001). However, some fellow hobbyists drew a further line of demarcation by refusing to eat their own animals and openly expressing disapproval for those who do. Similarly, within the commercial sphere, stockmen who worked with breeding animals tended to perceive this stage more favorably than did those who just buy in and finish store animals for slaughter. Unlike breeders, who often know their animals individually, finishers were thought to see livestock as mere "tools of the trade." Finally, older-generation farmers who took time to walk among and speak to their cattle were critical of farmers "nowadays" who check their beasts simply by driving jeeps and tractors through the fields. The older farmers saw this as a sign of increasing commercialization that reduces animals to just a means to an end. In sum, these types of productive differences become pivotal symbolic resources in terms of the narratives told and claims made by those engaged in such work.

Sociologists note that, when "people break social conventions and do something that appears immoral, strange, or untoward, they sometimes engage in motive talk to recast the meaning of their behaviour" (Arluke and Hafferty 1996, 201; see also Mills 1940). Although byre-face and slaughter-face workers might perceive themselves as "good people *doing* dirty work" (Davis 1984, 233), many face the challenge of giving accounts, to themselves and others, that not only justify or excuse their involvement in this line of human–animal work, but also neutralize any deviance associated with it (Scott and Lyman 1968; Sykes and Matza 1957). As C. Wright Mills (1940, 910) suggests, "What is reason for one man is rationalization for another. The variable is the accepted vocabulary of motives, the ultimates of discourse, of each man's dominant group about whose opinion he cares." This alerts us to how different cultural and emotional norms govern the actions and motives of different groups. The livestock sector is no different.

Animals and humans have long been valued and ranked in terms of how well they approximate the roles expected of them in society. This socio-moral placing of human and non-human animals constitutes "the sociozoologic scale" (Arluke and Sanders 1996, 168–169). As "good animals," pets and livestock enjoy an exalted servile status, albeit for different reasons. While pets are affectionately valued for their role as human companions, livestock are instrumentally valued as sources of food. Being little more than tools, food animals are, to all intents and purposes, "deanthropomorphized, becoming lesser beings or objects that think few thoughts, feel only the most primitive emotions, and experience little pain" (Arluke and Sanders 1996, 173). Ideally, people learn to relate differently to animals classified as "pets" and "livestock." For instance, it is socially acceptable to display emotional attachment to pets, but it is more appropriate to adopt "affective neutrality," or emotional detachment, toward livestock (Talcott Parsons, cited in Smith and Kleinman 1989, 56). Thus, someone who appears callous to his or her pet or overly involved with his or her livestock would be deviating from mainstream human–domestic animal feeling scripts.

In practice, however, people can be unsure how to categorize livestock and can hold mixed feelings toward them.[2] A good example of such ambiguity and ambivalence is the orphan pet lamb or calf scenario in commercial-productive contexts. Caring for orphaned or sick lambs not only offers farm children a relatively safe and formative experience; it also plays a key part in socializing them into the commercial pragmatics of producer–livestock relations. Even though youngsters will form varying degrees of attachment to such animals, they will learn that stock pets, at the end of the day, are working animals. Unlike real pets, productive pets will be sent to market or slaughtered.[3] Unsurprisingly, this can be a very poignant and traumatic experience. However, it also provides a pivotal lesson in "emotion work": "the act of trying to change in degree or quality an emotion or feeling" (Hochschild 1979, 561). If up-and-coming byre-face workers want to demonstrate appropriate feeling norms in keeping with commercial-productive contexts, and avert the emotional angst associated with pet-type relations, they have to learn strategies that foster a more reserved attitude toward the animals with which they work. Otherwise, fellow workers may ask them to account for excessive and superfluous displays of affection, a form of "rule reminder," or even sanction them by teasing. These types of informal exchanges effectively prompt those who deviate from the norm as to how they "should feel"—or, in this case, how they should *not* feel (Hochschild 1979, 564). Thus, emotion work "involves attempts to align privately felt emotions with normative expectations or to bring the outward expression of emotion in line with them." Arlie Hochschild (1983) refers to the first process as "deep acting" and the second as "surface acting," aiming to convey the fact that the first involves an attempt to change what is privately felt, while the second focuses on what is publicly displayed (Wharton 2009, 149). However, people-focused jobs such as those in the service sector have gone a step further and turned the personal management of emotion into a prescribed occupational requirement, or "emotional labor." For example, flight attendants, a typically female-dominated workforce, are trained "to create a publicly observable facial and bodily display" appropriate to serving the needs and expectations of the flying public (Hochschild 1983, 7).

In this case, agricultural workers, who are mainly men, are working not with people but with animals. From a young age they have been discouraged from actively forming relationships with, and having feelings for, their animals. In theory, this makes sense. The more livestock can be consistently perceived as commodities, the easier it becomes for agricultural workers to justify their actions to themselves, although not necessarily to others, which lessens the need for emotional management. This involves what Stanley Milgram has called "counteranthropomorphism," which is the "attribution of inanimate qualities to living things" (quoted in Arluke 1988, 100). The more livestock are de-animalized, the more this downplays, and possibly even denies, the animal's sentient nature. Objectification is an effective way to assuage any cognitive, moral, or emotional unease handlers might experience in the course of their work. And, of course, adopting an instrumental attitude is functional for the commercial sector because the workforce can efficiently process large numbers of de-anthropomorphized animals through the productive-slaughter system.

However, the commodity status of livestock, especially at the byre face, can be quite unstable. In practice, pursuing a thoroughgoing counter-anthropomorphizing strategy is inherently flawed, given that "animate nature can never be defeated totally; it still has a will" (Arluke 1988, 106). In other words, those who literally come face to face with livestock, especially bovine animals, soon realize that the commodities they handle can be unpredictable and demand respect. When animals cease to be mere livestock, and come to be seen as living beings in their own right, they potentially disrupt instrumental attitudes. Perhaps ironically, the point of slaughter can be a key time when byre-face and slaughter-face workers experience such disruption: "In the backdrop of death there are signs of life" (Wilkie 2006, 29–30). However, the more people stray from the tool-like status of livestock, the greater ambivalence and ambiguity they are likely to experience. In turn, this can make it trickier to justify their role, in terms of producing healthy animals for slaughter, and increases the likelihood of engaging in varying degrees of emotion work. The fact that most workers enjoy working with animals is just one of the core productive paradoxes they face. For some, it is simply more meaningful, and functional, to come face to face with some of the animals they work with than none at all, even though this can be more emotionally demanding. Perhaps having a few favored animals in the herd also gives vent to more socially acceptable ways to interact with animals, which partially offsets the more distanced approach in which they are expected to engage (Arluke 1988).

So far, the onus has been on stockmen to calibrate and display appropriate feelings in line with the productive context(s) in which they work. However, is the commercial sector on the verge of a transition that will shift the emphasis away from individualized strategies of emotion management toward a more institutionalized and feminized form of emotional labor? What follows is indeed speculative. But in light of recent changes, I suggest the established human–livestock interactional order is increasingly precarious.

When livestock were reclassified in European law as "sentient beings," they relinquished their de-animated status as "agricultural goods." Technically, this destabilizes the taken-for-granted tool-like status of food animals. Of course, the extent to which this reanimated classification will inform common-sense perceptions of livestock remains to be seen. However, if taken in conjunction with growing consumer demands to upgrade the welfare and productive contexts in which food animals live and die, these events indicate fundamental adjustments to how farm animals are legally, morally, and publicly perceived. At a practical level, the sector is also facing a human–animal skills gap, because the reservoir of skilled stockmen is drying up. This means that new recruits are less likely to come from a farming background and will have little, if any, experience working with cattle and sheep. By implication, novice workers are unlikely to perceive livestock as commodities, as they have bypassed informal farm-animal socialization processes that would encourage this instrumental attitude. Moreover, if more women are employed, as they are thought to be more caring, patient, and empathetic than male livestock handlers, this may feminize the cultural milieu of livestock markets, veterinary practice, and related animal-productive contexts.[4]

These changes potentially undermine the emotional distance, the feeling of aloofness from stock, that commercial-sector workers normally maintain and display. As productive benefits associated with and arising from "good stockmanship" are increasingly valued, perhaps the importance of empathy, a feminized attribute, will also come to the fore. To date, "The impact of the emotional relationship between stockperson and animal" has tended to play second fiddle to more scientific aspects of livestock management (English et al. 1992, 35). In an increasingly competitive global market, the commercial sector remains alert to new ways to manipulate, harness, and exploit every ounce of the animals it produces, and those who produce them. If cultivating human–animal rapport is shown to enhance animal productivity, this could become an occupational requirement and the cornerstone of good husbandry practice. Pursuing such a policy would also give the industry an opportunity to enhance its public image and animal-welfare credentials and to increase profits. But what may the emotional and psychological costs be for those compelled to engage in such labor? Clearly, that depends on which stage of the productive process it is applied to. If directed at the breeding end, it basically formalizes what more or less occurs unofficially. If extended to store animals, it would breach the aloof nature of human–livestock relations that typically characterizes this stage of the process. Experienced byre-face workers can sometimes be vexed sending animals to slaughter they have minimal or no rapport with, so how might inexperienced workers, formally instructed in the principles and practices of empathetic stockmanship, manage this situation emotionally? Given this unsettled backdrop, animal-farming norms that typically fashioned this productive interface can no longer be assumed.

But not everyone who works with livestock is exposed to the norms of mainstream farming—in this case, the masculinized world of beef-cattle production. For example, my hobby contacts were mainly novice female farmers who worked with sheep and functioned on the edge of the commercial sector. Many treated their animals virtually as outdoor pets, which points toward a less conventional set of producer–livestock "feeling rules" (Hochschild 1983). Such rules "describe societal norms about the appropriate type and amount of feeling that should be experienced in a particular situation" (Wharton 2009, 148–149). By imputing to farm animals "human-like 'feelings,' perceptions, sensitivities and even 'thoughts,'" these hobbyists drew on what Michael Lynch (1988, 267) has called the "naturalistic animal," which refers to common-sense perceptions of animals in everyday life. This clearly runs counter to the de-anthropomorphized, tool-like depiction of livestock described in the socio-zoologic scale.

Young children, workers from non-farming backgrounds, and hobby farmers seemed less likely to see livestock as tools of the trade, but were more likely to draw on lay understandings of animals.[5] To what extent do commercial farmers and stockmen, who have been born and bred into livestock farming, resort to this lay notion of animals? Just as scientists experimenting on animals have been inculcated into the norms governing their roles in the research community, as members of wider society "they [scientists] have inherited a highly complex but also highly ambivalent set of attitudes and beliefs about animals that tell them that they should respect animals and at the same time be able to use them" (Birke et al. 2007, 70). Officially, laboratory animals are presented as a means to acquire scientific data and as the "heroes of medicine" (Birke et al. 2007, 58). Informally, laboratory workers, especially technicians, may perceive and relate to some experimental animals in a pet-like manner. However, by the time research procedures are written up and published in scientific journals, any residual trace of the lab animals' status as pets has been distilled out. The accounts are impassive, and the notion of "analytic animals" has come to the fore (Birke et al. 2007). The analytic animal "is a *rendering* or transformation of the 'naturalistic animal' that relies upon commonsense access to laboratory animals while at the same time it rules out such informal modes of reasoning from relevance to the final product of the work" (Lynch 1988, 269).

I suggest that farmers and stockmen undergo a parallel dual apprenticeship of having to acquire and negotiate contradictory lay-agricultural attitudes to livestock. They, too, are familiar with everyday understandings of animals. But they have also been exposed to a secondary socialization process that encourages them to regard livestock as commodities.[6] In practice, those who handle beasts daily are most likely to revert to common-sense perceptions, especially if they work in breeding or showing or with sick or young animals, or if they identify with a few animals out of a herd. However, any animal, regardless of its productive function, that deviates from the routine process of production has the potential to stand out from the rest and become more than "just an animal." During the course of a normal day stockmen, oscillate, albeit to varying degrees, between instrumental and affective attitudes. A combination of "intrinsic" factors—that is, characteristics relating to the animal—and "extrinsic" factors, such as age, gender, and level of experience, seem to influence workers' attitudes toward the animals with which they work. Such factors have been called "attitude modifiers" as they have the potential to induce changes in "(a) people's affective/emotional responses to animals, and/or (b) their perceptions of an animal's utility to humans" (Serpell 2004, 147).[7] This points to the multifaceted, variable, and challenging nature of their role as carers for and economic producers of livestock. However, just as lay notions of laboratory animals are factored out and deemed irrelevant as those animals metamorphose into "analytic animals," as food animals approach slaughter, all traces of the naturalistic animal are also increasingly expunged as they are converted into "commodity animals."

Generally, agricultural workers—and, to a lesser degree, hobby farmers—seem to "need to compromise their sentimental attachment to animals with practicality" (Fukuda 1996, 16). In commercial contexts, where open acknowledgment of feelings is less acceptable, workers are encouraged to internalize any inappropriate feelings they might have. This is analogous to accounts of medical students who speak of "putting feelings aside" as they learn to touch live and dead human bodies, the "real work" of medical school (Smith and Kleinman 1989, 59). Moreover, in the course of everyday work life, it is unlikely that stockmen consciously think about concepts such as "pet" and "livestock" because their use of such categorizations "is largely unreflecting and pragmatic. Indeed, preoccupied with acting on the world in the 'here and now' they cannot afford to reflect on the adequacy of their type constructs and will only do so if the world starts to run counter to expectations" (Heritage 1984, 52). In addition, John Heritage suggests, "Each actor comes upon the domain of objects with different practical purposes in hand and knows that, 'motivationally speaking' they may be viewing the domain of objects in differently 'interested ways'" (Heritage 1984, 55). This point is important because how people view livestock depends on the extent to which it is primarily regarded as a commodity, and that varies according to a range of factors, such as the type and scale of hobby or commercial production, the stages within it, and where the person and the animal are positioned within that system. People will differ in their views. Indeed, one person may even hold manifold and irreconcilable views of the same animal (Humphrey 1995, 478). One way to illustrate this is through the typology of human–livestock relations set out in Table 9.1.

These four main patterns provide a snapshot of people's affective relations with different categories of livestock produced in hobby and commercial contexts. In theory, any livestock animal could be regarded as a pet. In my study, most pet-like animals are productive animals such as breeding cows and dairy goats. Hobby farmers are most likely to treat their animals in a pet-like manner and may even have animals that remain unproductive

TABLE 9.1 TYPOLOGY OF EMOTIONAL RELATIONSHIPS WITH LIVESTOCK

STOCK PET (ANY STAGE) *Strong emotional attachment*	PRODUCTIVE ANIMAL (BREEDING) *Emotional affinity*
• Pet-like status • Mainly hobby • Natural life span; may be buried • Slaughter still possible	• Individual status; "get to know" animal • Mainly commercial • Lengthened life span; remains productive • Slaughter after some years
CONSUMPTIVE ANIMAL (STORE) *Emotional aloofness*	COMMODITY ANIMAL (PRIME) *Emotional detachment*
• De-individuated status; part of group/batch • Commercial and hobby • Truncated life span • Fattened for slaughter	• "Tool-like" status • Mainly commercial • End of shortened/(re)productive life span • Slaughter imminent

throughout their natural life spans. In the commercial sector, however, stock pets such as orphan lambs routinely go to market or slaughter. Consumptive animals refer to store animals fattened for slaughter, as opposed to animals that produce products while alive, such as future stock, milk, and eggs. Unlike productive animals, which may live for ten years or more, consumptive animals have a shortened life span. For example, in my study cattle were killed by thirty months of age to enter the human food chain. Commodity animals are the finished product—in this case, prime animals slaughtered for meat. But this category also applies to productive-type animals that have come to the end of their (re)productive lives. However, having indicated the productive context in which these types of relationships are most likely to occur does not mean they could not occur in unmentioned contexts. Finally, it is important to note that the degree and kind of emotional involvement with each type of animal is not static and can vary as the animal's perceived status changes and the nature of the interspecies relationship strengthens or weakens.

For example, a commercial stockman working with a resident breeding herd of cows will have varying degrees of affinity to the animals with which he works. Out of the herd, a few cows may acquire a pet-like status, whereby the level of attachment increases, and a few may be more difficult to handle, which elicits a more aloof response. "Good" breeding animals tend to have a longer life span than those deemed to be less productive. Occasionally, a cow may remain on a farm even though its level of productivity has waned because it once produced good calves and the farmer has become especially fond of the animal. A new stockman, however, is more likely to send the animal for slaughter. One drawback of de-commodifying breeding animals is the potential unease of re-commodifying them when their reproductive lives are over. On paper, the process of transforming their status into commodity animals appears fairly clear-cut; however, the associated emotional transition can be tricky in practice (Wilkie 2005). Hobby farmers who re-commodify pet stock are faced with a similar challenge, but the level of emotional trauma they experience can be more tangible and intense in nature. "Good lifers" sometimes opt to bury favored animals on their land rather than send them for slaughter; such animals remain de-commodified. Finally, all producers try to detach themselves from store animals, but this can be especially difficult for hobbyists, who keep fewer animals and are more likely to know some, or all, of their animals more or less individually.

To varying degrees, all those at the byre face appear to negotiate and renegotiate the nature of their interactions with, and calibrate their feelings toward, the different types of animals they work with each day. Some people specialize in one stage of the process (e.g., breeding or finishing animals), while others work in breeder-feeder units in which they take newly born calves through the entire process to the point of slaughter. Producers might form varying relationships to animals within each of these categories and may even hold contradictory attitudes toward the same animal. However, not all byre-face workers are directly involved in the everyday responsibilities associated with looking after livestock. Mart workers and veterinarians work with a more transient population of unknown animals. This clearly throws up a different set of animal-management and-handling challenges. But even these workers can build up varying degrees of rapport with some of the animals they process and treat. Then there are dealers and estate managers, who tend to be the farthest removed from the daily responsibilities of the byre-face. Such personnel are most likely to view livestock as commodities because they have minimal, if any, direct contact with the animals. Even though livestock animals are in effect commodified sentient beings, in practice their status as commodities can be relatively unstable. This reminds us that commoditization is a process as opposed to a fixed state (Kopytoff 1986). For this reason, I use the term "sentient commodity" to draw attention to the ambiguous and dynamic status of livestock and the fine perceptual line workers have to negotiate in terms of seeing animals as both economic commodities and sentient beings.

I also suggest that stock pets and productive animals, whose career paths require a moderately elongated life course, are most likely to be actively recognized as individuals by those who manage them.[8] As such, they are more commonly named, are spoken to more frequently, are handled more positively, receive more veterinary attention, and elicit varying levels of attachment from their human supervisors. In contrast, consumptive and commodity animals have a short-lived career that necessitates a truncated and de-individualized life span, usually as part of a herd. This means that producers have less opportunity, and less time, to get to know them as individuals or form an affectionate relationship to them. Be that as it may, many agricultural workers and hobbyists also appear to "hold back" from actively forming bonds with store and prime animals because they are destined for slaughter. This does not mean that finishers do not know any of their animals. They just know fewer of them, as the majority will be processed as part of a faceless or anonymous herd. Then again, any animal that deviates from the routine process of production can come to the fore and become meaningful to those who work closely with them. For instance, this might apply to ill animals that require extra attention and physical handling and high-quality animals that stand out because of their exceptionally good conformation.

Clearly, livestock animals are not all the same; nor are the people who look after them or the productive contexts in which they work. One way to begin to comprehend the different types of emotional connection and disengagement workers might have with the different species and breeds of livestock they manage is to identify whether the animal is on a productive career path, and thus a life course, or a consumptive career path, and thus a death course, and then consider the extent to which, and under what circumstances, the worker's career path intersects with that of the animal. I also suggest that the attitudes and behavior of those working with livestock are bound up with the location of both humans and animals in the commercial and non-commercial division of labor. Factoring in all of these productive differences, as well as the scale and degree and type of intensive-extensive approach, provides a more nuanced, multifaceted, and flexible understanding of this rather neglected interspecies relationship.[9]

The instrumental and affective role of byre-face workers means they can face a range of emotional, cognitive, and ethical challenges that typically have been factored out of livestock-handling studies. Pursuing too narrow an understanding of the stockmanship role has meant that non-productive components of that role are little understood and underappreciated. Sociologically informed studies are especially well placed to address some of these gaps. For instance, exploring the experiences of those involved in breeding, feeding, marketing, medically treating, and slaughtering livestock—in this case, mainly cattle and sheep—has brought to the fore the gendered and ambiguous nature of animal-productive and animal-slaughter contexts. I have also shown that emotional work is associated not only with people-focused work; it is also evident in interspecies-focused work. Despite the growing interest in emotions within the discipline (see, e.g., Wharton 2009), mainstream sociologists involved in studies of work and organizations have yet to broaden their field to encompass non-human animals. As the livestock sector is on the cusp of undergoing fundamental changes in terms of having to attract non-farming men and women into the industry, this is an interesting time to conduct research into such settings. Moreover, the revised legal status of livestock could further undermine norms and rules about feelings that have long underpinned the commercial nature of human–livestock interactions. Thus, sociologically informed analyses could significantly complement and supplement findings derived from other academic disciplines, such as agriculture and animal science, to offer a more comprehensive appraisal of this challenging and changing interspecies work environment.

Appreciating the multifaceted nature of livestock production has implications beyond the academy too. As Harold Herzog (1993, 349) notes, "Disagreements about the treatment of animals in research ultimately stem from our tendency to think simply about complex problems." This seems relevant to the production of food animals, too. For instance, animal-rights and animal-welfare groups have raised the profile and plight of factory-farmed animals by unveiling the darker sides of these industries. The overall lack of productive transparency probably increased the impact of exposé investigations and further fueled public anger toward, and mistrust of, those involved in this much disputed sphere of animal production. The industry's apparent unwillingness to communicate openly with its customer base has also contributed to a perceptible lack of information from the perspectives of those who actually work with livestock. It thus seems that our current knowledge of productive-livestock relations is thus partial and somewhat partisan. Such gaps indicate blind spots and increase the likelihood of both sides' engaging in black-and-white thinking when discussing such matters.

Actors on both sides perhaps share more than they realize. Both are interested in food animals. While animal advocates, such as "new welfarists," are primarily concerned about the welfare of livestock animals, the industry as a whole is chiefly interested in animal productivity. Second, both groups have tended to under-appreciate and overlook the experiences and challenges faced by men and women who actually work with livestock and are ultimately responsible for ensuring the welfare and productivity of farm animals. In other words, I suggest that the actions, attitudes, and feelings of byre-face and slaughter-face workers are absolutely central to how livestock are managed and produced and should therefore be explicitly factored into lay and professional debates about food-animal production. Focusing on the animal side of the productive equation is clearly important, but it is only half of the story. We also need to gain a more comprehensive understanding of this contentious interspecies interface from the standpoints of productive practitioners.

Nowadays, most of us have little, if any, direct experience in livestock farming, so we are reliant on second-hand sources for our information. For instance, lay, academic, and media discussions frequently draw on animal-welfare and animal-rights-oriented perspectives and literature about such issues. But such groups have mostly concentrated on, and been especially critical of, the mass-production of food animals in hyper-intensive contexts: factory farming. The factory farm has become the dominant model for understanding livestock production in contemporary industrialized societies and is associated with an immediately recognizable narrative that elicits a fairly common response: It is bad as are the people who work in it. But even within factory farms, there will most likely be variations in the way people view and interact with animals. The extent of these variances requires further investigation. Such studies might consider how different species of livestock are perceived by different workers, at various stages of the production process, to be absolute commodities, especially as each stage provides varying opportunities and constraints on the nature of people's interactions with the animals with which they work. Although my research does not address those who work with factory-farmed animals, it would be interesting to see whether future sociological studies find any evidence of ambiguity and ambivalence in hyper-productive settings.

Animal advocates' accounts of livestock production have been very significant, but such accounts do not represent all of the contexts in which different species of animals are produced, or the full range of human–livestock relations that may occur within these settings. In other words, significant numbers of livestock are produced in factory-farm settings, but they are produced in other contexts too, such as hobby farms; small-, medium-, and large-scale farms; and organic farms. However, seeing livestock production primarily through the lens of factory farming is unhelpful, because one size does not necessarily fit all. As already suggested, farm animals and those who manage them are not homogenous; nor are the productive contexts in which they work. Adopting a more sociologically informed perspective demands a more nuanced and differentiated understanding of productive contexts. This not only brings to the fore the potentially mutable and ambiguous nature of interspecies relations. It also makes it more difficult to make clear-cut statements about such settings and those working in them.

Perhaps it is timely that a sociological dialogue is opening up around human–livestock relations, as it provides an opportunity to critically reflect, refine, and perhaps even dismantle some of the more intransigent discourses deployed by actors on both sides of this controversial interspecies issue. In my research, the accounts and experiences of those involved in small- to medium-scale productive contexts—and, to some degree, large-scale ones—indicate that livestock can be more than just "walking larders." Their ascribed tool-like status on the socio-zoologic scale is too narrow and too simple and does not represent the range of perceptions to be found among such agricultural workers. This book has focused on human–bovine relations and, to a much lesser extent, human–ovine relations. Future studies could consider how such findings compare and contrast with the other "big-three" food species: pigs, chickens, and goats. The global demand for meat and other animal-derived products shows no signs of abating. The wider implications associated with what has been dubbed the "Livestock Revolution" remains to unfold. Food animals are not new, but the global challenges ahead are. The legacy of animal domestication is as significant today as it was 10,000 years ago. We are entering relatively uncharted waters in terms of understanding the pragmatic nature of people's relations with livestock in modern industrialized societies. However, bringing to the fore the kinds of productive paradoxes and ambiguities that can underpin human–livestock interactions provides a fresh perspective on a long-standing interspecies connection. In the midst of uncertainty, "the troubled middle is not a comfortable place to be. But, for most of us, neither are the alternatives" (Herzog 1993, 349).[10] Irrespective of the rights and wrongs of rearing, killing, and eating "food with a face," one thing is for sure: Livestock will become deadstock for many years to come. We have an opportunity to more fully comprehend the challenges faced by humans and animals in a wide range of productive contexts. However, the extent to which we pursue this opportunity remains to be seen.

Notes

CHAPTER 1

1. In this chapter, I provide a historical overview and ethnographic account of the live export trade in the United Kingdom. I also explore the role of the organization Compassion in World Farming in raising the public profile of this issue.

2. This system is also referred to as "industrial farm-animal production" (Pew Commission on Industrial Farm Animal Production 2008).

3. The Pew Commission report (Pew Commission on Industrial Farm Animal Production 2008, 5) explores the use of antibiotics in livestock production and notes that it is difficult to assess the extent to which their use has contributed to resistant infections in humans. Nonetheless, "It can be assumed that the wider the use of antimicrobials, the greater the chance for the development of resistance."

4. BSE has been found in thirty-four countries other than the United Kingdom, including Canada, France, Italy, Japan, Spain, and, most recently, the United States (Nierenberg 2005, 41).

5. Although the prion theory is favored, it is disputed. Another hypothesis suggests that the agent could be a virus, because the BSE agent can "form multiple strains" (World Health Organization 2002). Another variation of the disease, called bovine amyloidotic spongiform encephalopathy (BASE), has been found in Italy (see Department for Environment, Food, and Rural Affairs 2007a).

6. Rendering describes how waste materials are processed to separate protein from fat (Smith and Bradley 2003, 187).

7. A ruminant animal grazes on grasses, herbs, and leaves. The rumen is "a fermentation chamber wherein an enormous population of micro-organisms works to digest cellulose and other plant fibres to yield nutrients" (Webster 2005, 130).

8. The effectiveness of this intervention would take about four years to assess because BSE appears to have a five-year incubation period (Smith and Bradley 2003). Others suggest that two to eight years might elapse before any clinical signs of the disease manifest (Kahrs 2004, 60).

9. Creutzfeldt-Jakob disease is a "chronic, fatal, degenerative neurologic disorder that occurs sporadically worldwide in approximately 1:1,000,000 people. Until 1993 it was regarded as a genetic, or metabolic, defect" (Kahrs 2004, 61).

10. Given the low incidence of BSE nowadays, moves are afoot to "increase, from 30 months to 48 months, the age at which cattle slaughtered for human consumption are BSE tested" (Food Standards Agency 2008).

11. In 1996, the entire head of a bovine animal, except its tongue, was added to the ban. "Since the banned tissues now contain more than offal, the tissues are referred to as specified bovine material" (Pattison 1998, 392).

12. See also Statutory Instrument 1995.

13. Some beef-suckler producers calve their animals in the winter. Calves run the risk of contracting enteric disease (i.e., diarrhea) if born inside. This can be very serious because, unlike barley beef calves that are artificially fed, suckler calves feed from their mothers. This makes it more difficult to replenish electrolytes to treat an animal's dehydrated state (Webster 2005, 151).

14. Some beef-suckler units might intensively finish the animals inside on concrete slats or in cattle courts. Barley beef systems also fatten dairy calves intensively inside on artificial diets, as they are separated from their mothers soon after birth. For more information about different beef-production systems and feeding regimes in the United Kingdom, see Red Meat Industry Forum 2007.

15. Demand for animal-derived produce in established industrialized countries is falling off and is anticipated to grow at a much slower rate (Steinfeld 2004, 20).

16. John Webster (2005, 252) prefers the term "well-being" to define "the state of being fit and feeling good since 'welfare' is used in common parlance in reference to a state of body and mind that may vary over the entire spectrum from healthy and happy to sick and desperate." He also draws on the animal-welfare principles set out in the "Five Freedoms" (discussed in Chapter 2) and a practical ethical approach called the "Ethical Matrix" put forward by Ben Mepham. This provides a framework to assess the costs and benefits associated with different productive approaches. It tries to take account of the various interests involved in food-animal production—that is, the different species of animals being produced; the producers; the consumers; and the environment.

17. For a critical discussion of terminology used in animal-agriculture settings, see Dunayer 2001. Joan Dunayer has created a thesaurus listing "alternatives to speciesist terms" in such settings. For example, instead of "livestock," she suggests, "enslaved nonhuman mammals," "captive nonhuman mammals," "exploited nonhuman mammals," "mammals enslaved for food," and "mammals exploited for food" (Dunayer 2001, 196).

18. A common assumption within the industry is that a productive animal is not suffering. This assumption has been undermined by a government report concluding that the rate of growth and level of productivity are unreliable measures of animal suffering because "growth, on occasion, can be a pathological symptom" (Brambell 1965, 10–11).

19. See, e.g., Eisnitz 1997; Gellatley 1996; Mason and Finelli 2006; Mason and Singer 1980; Nibert 2002; Penman 1996; Schlosser 2002; Singer 1995; Smith 2000; Tansey and D'Silva 1999.

20. Such animals are sent directly for slaughter because they are uniformly finished to pre-specified weight contracts. They are also more susceptible to infection; hence, the need for bio-security measures. For example, bypassing the mart would reduce the risk of contracting any disease, and when vets visit intensive pig units, they have to shower and change out of their outside clothing.

21. I attended sales and shows at the auction market, shadowed auctioneers and mart workers, shadowed vets in a mixed veterinary practice for a month, was shown around a specialist abattoir that slaughtered cattle, and attended a rare- and minority-breeds livestock sale and open day. Moreover, I conducted seven interviews with either two or three people present. The additional people were work colleagues, spouses, or members of the interviewee's family.

22. There were thirty men and twenty-three women. I interviewed thirteen farmers and stockmen, eighteen mart workers and auctioneers, twelve hobby farmers, five veterinary workers, and four abattoir workers. For a more detailed biographical summary of my contacts, see Wilkie 2005. I also designed a profile questionnaire for interviewees to complete that elicited biographical details and asked a few open- and closed-ended questions relating to their experience with domesticated animals (livestock and pets). The spouse of one contact did not return the sheet, so all percentages are based on those who did ($N = 52$).

23. Schoolteachers work with pupils at the "chalk face." I use the term "byre face" to describe people who directly interact with livestock as part of their everyday work lives. "Byre" is a Doric word for cowshed, and Doric is the vernacular of rural communities in northeastern Scotland (Kynoch 1996).

CHAPTER 2

1. The terms "stockmanship" and "stockman" tend to apply to men but can refer to women too (Farm Animal Welfare Council 2007, 2). I have used these terms on this basis, too.

2. See Ingold 2000 for an anthropological critique of this assumption.

3. Until the seventeenth century, the term "savage" referred to plants. It was then extended to describe "primitive" peoples (Anderson 1997, 474).

4. Women's ability to give birth reinforces the idea they are more embodied and closer to nature than men (Ortner 1974).

5. Cultivating plants attracted wild herds of sheep and goats to eat these crops. This eased the hunting of such animals, but the persistent "crop-robbing" significantly depleted food earmarked for winter months (Swabe 1999, 31). This may have induced people to domesticate animals as a strategy to manage their raids: "Large ruminants came as crop-robbers and ended up as domesticated beasts" (Zeuner 1963b, 10).

6. The Near East is described as the "hearth of agriculture" (Simon Davis, cited in Swabe 1999, 30). The Fertile Crescent refers to areas of Turkey and Iran that bordered northern Syria and Iraq (Toussaint-Samat 1994, 95).

7. Other domesticated species include "water buffaloes, yaks, horses, asses, camels, llamas, reindeers, ducks, geese and turkeys" (International Livestock Research Institute 2007, 24).

8. For Frederick Zeuner (1963b, 9), domestication is a "biological phenomenon" because "the art of domestication is not restricted to the species *Homo sapiens*. There are, for instance, many species of insects that domesticate other insects, and if one looks at this phenomenon in an unbiased way, one finds that it does not differ from that observed in the case of man and the various animals and birds he has domesticated."

9. The cat is an exception to this rule. "Taming" refers to "a relationship between an individual animal and an individual human whereas 'domestication' involves populations and successive generations" (Russell 2007, 31–32). It is thought that the rearing and taming of young animals also contributed to the domestication process (Swabe 1999; see also Serpell 1989, 1996). See Budiansky 1992 for a critique of this thesis.

10. An animal's flight distance describes the point at which it flees from a person that approaches it (Hemsworth and Coleman 1998, 41).

11. See Franklin 2007; Harvey 2007; Twine 2007.

12. The Chesapeake area includes Virginia and Maryland. The New England area includes Connecticut and Massachusetts.

13. In the late eighteenth century, there was an influx of Cheviots, an improved breed of sheep, into the Scottish Highlands. This led to the Highland clearances, in which about 15,000 Highlanders were evicted from their leased homes and land by wealthy landowners who saw an opportunity to make money farming sheep (Franklin 2007, 108–113).

14. This period saw the mass slaughter of bison herds from the Great Plains. I discuss this further in Chapter 5.

15. Aberdeen Angus were first imported to America in 1873: see also Walton 1999.

16. Longhorns are immune to Spanish fever, but imported cattle breeds did contract the disease (Carlson 2001).

17. By the mid-1880s, Arizona, Colorado, Kansas, and New Mexico had instituted quarantine laws that prohibited cattle from the South from entering or passing through these states. Large-scale ranchers fared better than their small-scale counterparts in absorbing the costs of mandatory requirements to eradicate ticks (Carlson 2001, 83–103).

18. By the 1880s, "native Indian" populations had been killed or subdued through a policy of starvation. The extermination of bison paved the way for "strange beasts" but removed their means of subsistence. Indigenous peoples forced onto reservations depended on meat from livestock, and competition for reservation beef contracts was fierce because they were so lucrative.

19. For a contemporary description of cattle being finished in a feedlot, see Pollan 2006.

20. By the nineteenth century, a "democratization of meat" had occurred, as the lack of meat in a working class diet "was no longer seen as normal, but increasingly as a form of deprivation" (Knapp 1997, 544).

21. Swift was one of five main meatpacking companies that formed the so-called "Beef Trust." As these companies cornered the meat market, they could under-cut their competitors (Rifkin 1992, 114).

22. Ice stations were located next to rail networks.

23. This policy is discussed more fully in Chapter 5.

24. In 2006, the EU banned the dietary use of antibiotics as growth promoters (Union of Concerned Scientists 2006).

25. For a critical review of this work see, e.g., Waiblinger et al. 2006.

26. I have italicized Doric words in the text and given their meanings in brackets the first time they are used (see also the Glossary in this volume).

27. "Consubstantial" refers to how hill farmers, through their everyday practice, do not differentiate the (spatial) relationship that exists between their families and their farms: "The distinct existence and form of both partake of or become united in a common substance" (Gray 1998, 345).

28. I interviewed twenty-three women.

CHAPTER 3

1. This is the second verse of a well-known traditional children's song. For a full version of the lyrics, see the British Council Web site at http://www.britishcouncil.org (accessed August 27, 2009).

2. Since the 1980s, this gap in the literature has been steadily addressed: see, e.g., studies carried out in Australia (Liepins 2000); Canada (Leckie 1996a); Denmark (Pedersen and Kjærgård 2004); France (Saugeres 2002a, 2002b, 2002c); Ireland (Shortall 2001, 2002); Norway (Brandth 2002a, 2002b; Haughen and Brandth 1994); Scotland (Schwarz 2004); the United Kingdom (Gasson 1980; Little 2002; Morris and Evans 2001; Price and Evans 2006; Wallace et al. 1996; Whatmore 1991); and the United States (Maret 1993; Neth 1995; Rosenfeld 1985; Ross 1985; Sachs 1983; Trauger 2001, 2004).

3. A U.S. survey of farmwomen reinforces this point: "Women on farms selling mainly animals (in whose production women tend to be active) were involved in a somewhat wider range of decisions" (Rosenfeld 1985, 274).

4. Ulrike Schwarz's (2004) study of ten medium-size family farms in Southeast Scotland explores how boys and girls are socialized into, and show a preference for, a wide range of farming-related tasks.

5. Even though it was acceptable for women to kill chickens, animals such as "kittens to cattle" ought to be killed by men (Hunter and Riney-Kehrberg 2002, 141).

6. This applied equally to dairying in Canada, in the United States, in all of Ireland (i.e., the historical period under review was pre-partition), and in Denmark (Shortall 2000, 247).

7. Perhaps different dairying regions within a country would undergo change more or less quickly. For example, farm girls in Devon, England, were still being trained as milkmaids in 1913 (Ley 1978).

8. See, e.g., Food and Agriculture Organization (1995, 2000) for insight into the diverse nature of women's roles and level of involvement in non-commercialized livestock farming systems in developing countries.

9. This may also depend on whether different geographical regions are more or less associated with producing different species of livestock.

10. The spouse of a former slaughter man was excluded from this summary because she had no experience with livestock.

11. Hobby farming and rare breeds of livestock are explored more fully in Chapter 5.

12. Some agricultural workers may have owned or run family farms, but I did not ascertain how many animals they owned because I focused on their experiences of working with livestock in the mart.

13. Dairy cows may be more likely to be located at the female end of this continuum.

14. Ryanne Pilgeram (2007, 582) also notes, "The dangerous, dirty job of ring man, complete with charging bulls and flying manure, was reserved for men."

15. The extent to which women are believed to have greater affinity to, and interconnections with, animals are explored in Adams and Donovan 1995; Gaard 1993.

16. "*Farmtoun*" or "*fermtoun*" refers to a farmhouse and its accompanying land and buildings (Cameron 1978).

CHAPTER 4

1. In 1746, rinderpest had a devastating impact on the cattle-droving trade.

2. Since dealers from the south were reluctant to embark on arduous treks to and from the Highlands, a new "class of hardy ànd adventurous men" came to the fore: the drovers (Haldane 2002, 46).

3. "The first public livestock auction in America occurred in Ohio in 1836. This sale by the Ohio Company was also the first sale at auction of purebred cattle in America" (Boeck 1990, 24–25). In the 1930s and 1940s, the number of auctions escalated throughout America, from 200 to about 2,000. It is believed that the auction system was carried by those who had migrated from England to America (Cassidy 1967, 33).

4. There are international variations in how auctions are conducted (Cassidy 1967). For example, the Dutch system is characterized by a descending pricing mechanism; the auctioneer initially kicks off the sale with a high price and gradually decreases the price until the first person places a bid.

5. Arthur Young, an agricultural reporter of the time, notes that Bakewell was a tenant farmer in 1770 (Trow-Smith 1959, 55).

6. According to the Collins English Dictionary and Thesaurus (Glasgow: Harper Collins, 1993, 1356), a yeoman is "a member of a class of small freeholders who cultivated their own land."

7. Bakewell "was by occupation purely a breeder of pedigree stock" (Trow-Smith 1959, 46).

8. Sheep were especially valued for their wool; for many English farmers, this was a key source of income.

9. Prized animals could significantly appreciate in monetary value, too.

10. The Markets and Sales Order (1903) made it mandatory for markets to wash their floors, to try and minimize, and even prevent the spread of animal disease (Perren 1978, 139).

11. A shilling is 5 pence in contemporary money, so 45 shillings would be about £2.25.

12. The Markets and Fairs (Weighing of Cattle) Act of 1926 made it compulsory to weigh cattle in markets and inform bidders of these weights (Brian Harvey and Frank Meissel, cited in Graham 1999, 47).

13. Auctioneers in Midwestern and western areas of the United States at times adorn cowboy hats and boots during sales. This draws on potent historical and traditional images "to authenticate their auctions" (Smith 1989, 115).

14. A "crush" is a metal crate used to restrain cattle; in this case, it is used to check the numbers on ear tags and dentition.

15. For an ethnographic and gendered account of attending a U.S. livestock auction in Oregon, see Pilgeram 2007.

16. Intensively produced pigs and poultry are more standardized than sheep and cattle. They also bypass the mart, this partly relates to their being more susceptible to infections. This was not always the case. As one auctioneer explained, pig and poultry sales were the biggest sales of livestock in the 1960s, but as these species underwent intensification, it increased contract specifications (e.g., standardized weights), which, in turn, led to more "uniform lots" of pigs and poultry.

17. Producers are open to an array of biological, climatic, and geographical factors that may have unexpected effects on the productive process, some of which they can control, and some of which they cannot.

18. Livestock shows that emerged at the end of the eighteenth century were known as cattle shows, "not because cattle were the only or even the main class of stock exhibited," but because the terms "cattle" and "livestock" were interchangeable (Fraser 1960, 51).

19. I explain this grading system in more detail later in the chapter.

20. This refers to a special beef premium that producers could claim for male cattle: two claims for steers (i.e., castrated males), and one claim for bulls. The scheme started in 1994 and ended in 2005. It is now a Single Farm Payment instead.

21. All cattle born in the United Kingdom after July 31, 1996, had to have a passport.

22. Pictorial examples of cattle ear tags and passports can be accessed through Department for Environment, Food, and Rural Affairs 2007b.

23. The Animal and Plant Health Inspection Service of the U.S. Department of Agriculture currently oversees a voluntary tracking system. However, moves are currently afoot possibly to extend animal registration and make it a mandatory requirement. The Pew report indicates that the British Cattle Movement Service system in the United Kingdom could be a suitable template (Pew Commission on Industrial Farm Animal Production 2008, 67).

24. This EU regulation took effect on December 31, 2009, and replaced the current identification and batch system. For more information about Electronic Identification (EID), see the Department for Environment Food and Rural Affairs Web site at http://www.defra.gov.uk (accessed December 21, 2009). However, the National Farmers Union was opposed to EID on the grounds that it was too costly and impractical (see National Farmers Union 2008a).

25. See Statutory Instrument 1998.

26. A mart worker said that farmers are less likely to separate cows from their calves when selling. Dealers are more likely to do so if they think they will get more money by selling the animals separately.

27. "The box" is a historical reference to the auctioneer's rostrum in Scotland (Thomson 2005, 30).

28. On the rise of electronic marketing systems in the United Kingdom, North America, and Australia, see Graham 1999.

29. This also applies to rare breeds of livestock.

30. The Meat and Livestock Commission was disbanded in 2008.

31. There are up to eight conformation classifications for cattle carcasses, as opposed

to five for sheep. The U, O, and P grades have been subdivided into higher and lower band ranges.

32. This applies to sheep carcasses. For cattle, it is bands 4 and 5 that are subdivided into L and H.

33. See Livestock Knowledge Transfer 2001, 315.

34. Genetic indicators such as estimated breeding values are now being used by breeders to assess the breeding and fat potential of their livestock. Thus, "Rather than publicly displayed ritual, evaluation is technical and private, requiring expert scientific and statistical knowledge practices to produce numbers supposedly representative of an interior genetic composition" (Holloway 2005, 895).

CHAPTER 5

1. In this chapter I refer to data and themes discussed in Wilkie 2006.

2. Hobby farming flourished in North America because it is a wealthy, "highly mobile" region where there was "a relative lack of planning control" (Layton 1980, 220).

3. Keith Halfacree (2007, 4–5) writes, "Back-to-the-land is a player in the game of producing rurality/ruralities for the new millennium. How important a player it turns out to be remains, of course, a matter rooted in politics, power, advocacy and imagination."

4. Alexander Mather and his colleagues explain that this figure was suggested by a high-profile estate agent who processes about one-fifth of farm properties advertised in *Farmers Weekly* and *Country Life*.

5. These figures apply to rural England.

6. Terms such as "hobby farming," "part-time farming," and "smallholder" are used more or less interchangeably in the literature, but different disciplines and institutions, such as agricultural, social science, and government researchers, use and define these terms in different ways. For further discussion, see, e.g., Boyd 1998; Gasson 1988, 1990; Holloway 2000a, 2000b, 2001, 2002.

7. *The Good Life* was a situation comedy shown on television in the United Kingdom between 1975 and 1978 (TV Comedy Resources 2008).

8. A wedder is a castrated ram.

9. Part-time farming has been perceived as "farming on the fringe" and not "'real' farming" (Nigel Robson, cited Gasson 1988, 1).

10. Walbert requests that his work be copyrighted 1997–2004.

11. A similar attitude has been expressed toward hobby-bison producers in America. The commercial sector is concerned that "hobbyists will damage their image by failing to adopt conventional management practices and selling" substandard meat produce. Given this, they are seen as "urbanites at best, and more critically as unhelpful outsiders" (Lulka 2008, 39).

12. Livestock conservation came to the fore at the Biosphere Conference in Paris, following an initiative taken by the United Nations Food and Agriculture Organization in 1966 (Alderson 1994).

13. For a detailed discussion of the criteria underpinning rare breeds, see Alderson 1993.

14. These categories apply to cattle, sheep, pigs, poultry, horses, and ponies, but "other native breeds" does not seem to apply to goats.

15. Breeds acknowledged by the RBST "must be an original breed, or a native breed of which at least one parent breed is believed to be extinct." Finally, an "original population" refers to the same criteria for an original breed "which has not suffered unacceptable introgression" (Rare Breeds Survival Trust 2008c).

16. In the 1970s, a small cohort of bison ranchers produced bison meat for specialist markets (Lulka 2008, 41).

17. The idea that "improved" commercial breeds are superior to unimproved breeds is evident in the arable side of farming, too. For example, researchers summarizing the emergence of the "Dutch wheat regime" note that seeds derived from this regime—the so-called improved varieties—quickly became the social norm in Dutch agriculture." Farmers who did not use improved seeds were described as backward, and their seeds were branded "inferior" (Wiskerke 2003, 438–440).

18. Following the Maastricht Treaty in 1992, the EEC became known as the European Union. There are currently twenty-seven member states.

19. The "post-productivist" concept is contested. For example, it is predominantly, although not exclusively "UK-centric" (Wilson 2001). Most discussions have applied it to agriculture and have paid less attention to non-agricultural land use, such as forestry. Others are keen to "focus on land use rather than on rural social change" (Mather et al. 2006, 441). Some query the viability of the term itself, and the prefix "post-," as it implies a transition from one productive situation to another (Evans et al. 2002). Others favor the concept of a "multifunctional agricultural regime" instead (Wilson 2001, 94; see also Wilson 2008).

20. David Lulka (2006, 173) prefers the concept of "hybrid agriculture" to the dualistic productivist–post-productivist schemata because it puts more emphasis on "the material and ethical importance of nonhuman agency in agricultural systems."

21. On the ethical implications of "discarding" breeds, see Jackson 1998.

22. There is much debate about cheap food because hidden costs, such as health, the environment, transportation, animal-welfare standards, low wages, and poor working conditions are not factored into the "true" price of such products: see, e.g., Lawrence 2004; Nerlich 2004; Lang and Heasman 2004; Stevenson 1997.

23. A similar rationale was proposed by the executive director of the National Bison Association in America: "I think that we are going to play a big role in building the breed but these would still be very exotic rare animals if it wasn't for the fact that people are eating them" (quoted in Lulka 2008, 50).

24. This scheme was set up in 1994. Since 2003, it has been overseen by the Traditional Breeds Meat Marketing Company, which authorizes "independent butchers to stock and market rare and traditional breeds" (Rare Breeds Survival Trust 2008a).

25. According to the RBST's Web site, there are none in Scotland or Northern Ireland: see http://www.rbst.org.uk/accredited-butchers (accessed July 14, 2009).

26. See Evans et al. 2002 for a critique of the assumption that post-productive agriculture is associated with high-quality food products.

27. In response to an article by Valerie Elliot (2008), the National Farmers Union issued a position statement claiming that the remarks of its president had been "taken out of context and their meaning distorted in order to create controversy" (National Farmers Union 2008b, 1). The alleged citation refers to whether hobby farmers "should be expected to have a licence or some other form of certification if they keep livestock, so as to reduce the risk their activities might pose to professional livestock farmers" (National Farmers Union 2008b, 1).

28. For concern about RBST members' flouting welfare legislation when transporting "pets," see Todhunter 1995.

29. The niche market for rare-breed meat also reflects a productive role: see Clutton 1991.

30. Nick Evans and Richard Yarwood (1998) are referring to data they collected in 1996 when they surveyed RBST members. Yarwood et al. 1997 includes owners from Northern Ireland and the Republic of Ireland. The questionnaire was sent to all Irish RBST members ($N = 131$); Evans and Yarwood received thirty-two completed schedules, giving them a 24 percent response rate. In addition, they received 111 questionnaires from people who lived in Britain who were keepers of Irish breeds. Unless stated otherwise, findings refer to the Irish respondents.

31. Very rare breeds tend to be located where the breed originated. For example, during the 1970s the last herd of Gloucester cattle was based in Gloucestershire (Yarwood and Evans 1998, 151).

32. See Yarwood and Evans 2000 for evidence of RBST members in Scotland, England, and Wales, respectively, owning "Scottish," "English," and "Welsh" rare breeds (see also Yarwood and Evans 2006).

33. Examples of rare-breed icons of historical locations are Cotswald sheep, which represent the Cotswald Area of Outstanding Natural Beauty; the Dartmoor Pony, the symbol of the Dartmoor National Park; and Swaledale sheep, which have become synonymous with the Yorkshire Dales National Park (Yarwood and Evans 1998, 151).

34. A rationale for preserving all rare breeds of livestock cannot always be found, because some are unable to fulfill any or all of the post-productivist roles. Despite this, some argue that "all endangered breeds need to be conserved as an insurance against unknown circumstances" (Dowling et al. 1994, 17).

35. "Soay" means "sheep island" (Alderson 1994, 49).

36. The RBST has more than 8,000 members (Evans and Yarwood 2000).

37. Evans and Yarwood 1998 notes that "keepers" were more likely to respond to the survey than "non-keepers."

38. According to 1991 Census of Agriculture, Statistics Canada, out of a total of 50,991 hobby farms, 30.8 percent kept cattle, 2.6 percent kept hogs, 2 percent kept sheep, and 1.3 percent kept poultry. There were .5 percent that kept cattle, pigs and sheep (Boyd 1998, 7). Unlike for the RBST hobby farmers in the United Kingdom, the preferred species in Canada appears to be cattle. No information was given to clarify whether the Canadian hobbyists kept commercial or rare breeds. Although the data are somewhat dated, they indicate potential differences.

39. Evans and Yarwood (1998) did not state how many men and women made up their sample. The sample in Holloway 2000b sample also was middle-aged. The demographic characteristics of Canadian part-time farmers in 1991 indicated that 26 percent were women, and the average age of all part-time farmers was 45.2 (Rick Harrison and Sylvain Cloutier, cited in Boyd 1998, 3).

40. Keepers could give more than one response in this section of the survey (Evans and Yarwood 1998, 10).

41. Horses were working animals, but with increasing mechanization, they were displaced and replaced by tractors. Moreover, in some countries, such as France, horses can be a source of meat.

CHAPTER 6

1. Before the nineteenth century, no European country had passed laws to protect animals (Maehle 1994, 95).

2. For historical overviews of people's indirect duty and obligations to animals, see, e.g., Maehle 1994; Thomas 1983.

3. Part I of the Agriculture (Miscellaneous Provisions) Act was passed in 1968. The act responded to recommendations put forward in Brambell 1965 and redefined suffering in terms of "unnecessary pain or unnecessary distress" (Harrison 1970, 10).

4. See also Radford 2001; Webster 1994. For a critical appraisal, see Buller and Morris 2003.

5. For a historical overview of different philosophical and scientific positions and perspectives pertaining to animal sentience from the Renaissance period to the current day, see, e.g., Duncan 2004, 2006; Turner and D'Silva 2006; Webster 1994, 2005.

6. A legal case in America (*Corso v. Crawford Dog and Cat Hospital Inc.* [1979]) ruled that pet animals "occupy an intermediary space between property and legal personhood" (Yancy 2006, 199).

7. During the 1990s, there was a linguistic and political shift away from "pet owners" toward "guardians" of "companion animals." The shift is animal-centric, as it acknowledges that animals "possess a level of consciousness that makes them similar to humans in many ways" (Irvine 2004, 60–61).

8. I corrected a printing error in the original source.

9. I have used these terms interchangeably to avoid overusing the term "livestock."

10. This statement was in my interviewee profile sheet.

11. Clearly, this is not always the case (Tuan 1984)—for example, animal abuse, neutering, obesity, and unwanted and problematic pets. Animal shelters attempt to find new homes for unwanted animals, but if unsuccessful, those animals may be euthanized (see, e.g., Arluke 2004, 2006; Irvine 2003). The "darker sides of pet keeping" have led some to argue this is an exploitative relationship too, and should be stopped (Serpell 1998a, 112; see also Bryant 1990).

12. According to a study that compared and contrasted attitudes held by urban and rural residents toward dogs and calves, Isabelle Veissier and Patrick Chambres found "little evidence of the extension of empathetic 'pet' style representations by urban residents to farm animals. On the contrary, the farmers interviewed by Veissier and Chambres were notable for the emphasis they gave to the non-material characteristics of individual farm animals" (cited in Buller and Morris 2003, 225–226).

13. If she was handling sheep or poultry, she may not have felt the need to "harden up" to the same degree, because those animals are perceived to be less intimidating and safer to handle than cattle.

14. I use the term "explicit" to highlight the fact that, legally, companion animals are commodities, too, but their status as such tends to be more implicit. Second, some productive animals may remain de-commodified and do not re-enter the production process. To all intents and purposes, they are perceived as pets and will die a natural death.

CHAPTER 7

1. This chapter draws on arguments and data I discuss elsewhere (Wilkie 2005).

2. The sample for the study was selected over three separate years. In 2000, Anne-Charlotte Dockès and Florence Kling-Eveillard identified a total of ninety livestock breeders (dairy [20], beef-suckler cattle [20], pigs [25], and poultry [25]). In 2001, they met thirty advisers from a range of farm organizations and services, including veterinarians, and in 2004, they focused on farmers affiliated with quality and animal-welfare assurance schemes. This applied to sixty producers: veal calves (15), pigs (15), dairy (15), and laying hens (15). More information about the sample in terms of age, gender, herd size, and commercial or hobby farming would have been useful to contextualize the findings.

3. Henry Buller and Carol Morris (2003, 233) make the point that "quality assurance" and "farm assurance" schemes are often spoken about as if they are the same thing. However, according to the U.K. Farm Animal Welfare Council, quality assurance is more about making sure consumers are reassured that predetermined standards associated with a productive process have been adhered to. Farm assurance is as an offshoot of the quality assurance, as it relates to the actual application of quality-assurance standards throughout the production process— that is, from farm to slaughter.

4. U.S. cattle ranchers also get to know some of their animals as individuals with their own personalities. They also harness this knowledge when physically handling or moving the animals (Ellis 2007).

5. He fattens sheep over the winter, too.

6. This may change in more automated or robotic milking systems, where there is less direct interaction with humans (see, e.g., Holloway 2007; Wathes 1994).

7. This also applies to fish.

8. Alarm or stress triggers a physiological "fight or flight" response, which stimulates the release of hormones such as cortisol and adrenaline from the adrenal cortex and medulla (Webster 2005, 28).

9. In practice it is unlikely that workers will emulate this ideal all of the time.

10. Commercial sheep farmers were not included in this particular study.

11. Recent research has explored additional factors that may influence the nature of farmers' relationships with their animals: the species being produced, the way animals are housed, stock density, life span of animals on a farm, and generational links to the farming family through livestock bloodlines and breeding (Bock et al. 2007).

12. A gilt is a young sow that has not given birth to her first litter of piglets.

CHAPTER 8

1. Muslim and Jewish methods of slaughtering livestock are two notable exceptions to the United Kingdom's stunning legislation: see, e.g., Burt 2006; Levinger 1979.

2. This comment influenced the title of the book.

3. I discuss this concept further in Chapter 9.

4. Vets can get to know some animals if they have prolonged contact with them or input into their care.

5. For examples of bad practice in slaughterhouses, see Eisnitz 1997; Smith 2000; Warrick 2001.

6. For a discussion of why slaughtering and butchering are male-dominated jobs, see Pringle and Collings 1993.

7. If they slaughter 600 cattle, they get a bonus to reach 1,000.

8. A motor drives the chain, and the foreman can adjust the speed. The system uses more employees and increases the throughput of animals being slaughtered. The chills refer to big refrigerators in which finished carcasses are stored for dispatch to butchers and supermarkets.

9. But see n. 1.

10. British vets "entered into public service and the realms of disease management" after the major cattle-plague epidemic of the mid-1860s (Hardy 2003, 3). The passing of the Public Health Act in 1875 opened slaughterhouses to external scrutiny, but the official inspection of meat was mainly carried out by medical professionals, not by vets. The extent of veterinary involvement generated "inter-professional tension" as human- and animal-health experts tussled over their disciplinary contributions and interests (Hardy 2003, 14). Moreover, Abigail Woods (2007) illustrates how pivotal the interwar years were in terms of the veterinary profession's demonstrating to skeptical farmers and government officials the valuable contributions it could make to livestock production—in this case, via cattle breeding. This not only secured the future input and expert status of vets in this sector, but it also meant that vets increasingly superseded farmers as experts in bovine reproduction.

11. Vets are reluctant to take meat-hygiene positions in the United Kingdom. Some think it is not "real" veterinary work or that it is for those who want to retire from "real" practice (Huey 1998, 44).

12. There are two stages to slaughtering cattle: stunning and sticking. The first stage is usually achieved using a captive bolt gun: "When the cartridge is fired by pulling the trigger, or by contact with the head of the animal, the bolt shoots forward, penetrates the skull and stuns the animal by concussion" (Webster 2005, 172–173). If it is carried out competently, it should leave the animal unconscious, so when the carotid artery and jugular vein in the animal's neck is cut to initiate the bleeding-out process, the animal should not experience pain or discomfort. The colloquial term "spring" refers to "the laceration, after stunning, of central nervous tissue by means of an elongated rod-shaped instrument introduced into the cranial cavity (pithing)." This intervention is believed to minimize powerful involuntary movement of an animal's limbs after stunning, which could injure a slaughter worker when shackling the animal's leg before sticking. Pithing also prevents animals from regaining consciousness after stunning. Because this practice can disperse BSE-contaminated brain tissue through an animal's body, recommendations were made in 2001 to stop it (Food Standards Agency 2001, 1).

13. Animal-rights proponents such as Tom Regan are against an unconditional approach to euthanasia, especially in situations where healthy animals are killed. Instead, a number of caveats should be satisfied before the act is performed. The method of killing should be as pain-free as possible, and the person who decides and the person who performs the act should both be convinced it is in the animal's interests (Tom Regan, quoted in Swabe 2000, 301).

14. Large-scale production systems can change the meaning and significance of animal death. For example, the different stages of intensively produced pigs are associ-

ated with varying levels of death; such losses are normalized and expected. Some suggest that "the fact that death on the farm has become an ordinary matter has brought about changes in representations of livestock farming work, transforming a life-bearing craft into death work" (Porcher 2006, 67–68).

15. A Danish study explored the reasons why dairy animals in small and large herds were being euthanized and found that animals in large herds were more likely to be slaughtered on the farm (this was the researchers' definition of euthanasia). The study found that the chances of euthanasia "increased with increasing herd size, increasing average herd milk yield and an increasing average number of disease recordings per cow per year" (Thomsen and Sørensen, 2009, 44). It is also interesting to note that, in the list of primary reasons stated by farmers, 8.4 percent of euthanized cows ($N = 787$) were recorded as "unknown." It may be worth exploring the reasons in the unknown category more fully because they could reveal *non-productive* reasons, such as the nature of the relationships farmers have with different cows in their herd.

16. Commenting on factory farming, where poultry farmers might "exterminate" slow hatchers thought to be weaker than fast hatchers, she suggests that the mass-production of chickens contributes to such animals' being seen as a "more expendable commodity than larger and more expensive forms of livestock" (Harrison 1964, 2–3).

17. When this scheme ended in July 1999, farmers faced the productive dilemma of rearing dairy-bred bull calves for an already saturated market or spending money to cover the costs of sending the animals to slaughter (Gatward 2001, 74).

18. See Woods 2004 for a historical analysis of the U.K. government's rationale to reject vaccination as an alternative approach to managing and controlling foot-and-mouth disease.

19. See Franklin 2007 for an analysis of U.K. newspaper coverage of foot-and-mouth disease. Franklin alludes to "Phoenix," a newly born calf that survived the culling process. The baby bovine celebrity captured the public's imagination and affection. Although Phoenix was destined to be killed, Prime Minister Tony Blair intervened and overruled the official cull policy.

CHAPTER 9

1. For a discussion of how American meat cutters and Turkish butchers find honor in the "dirty work" they do, see Meara 1974.

2. Ambiguity arises when we are unsure what to think about something, and ambivalence occurs when we experience mixed feelings (Weigert 1991, 17).

3. The fate of a laboratory pet is different from that of a productive pet because "the meaning of a pet in this setting is constructed in the shadow of sacrifice; the animal is chosen or elected to be spared death" (Arluke 1988, 106).

4. For evidence of increasing numbers of women practicing veterinary medicine in the United Kingdom and Canada, see Lofstedt 2003; Paul and Podberscek 2000; Robinson and Hooker 2006. For information about Lantra's initiative to attract more women into agriculture and other land-based industries, see Lantra 2009. Lantra is the Sector Skills Council for environmental and land-based industries.

5. Similarly, technicians, research fellows, and visitors who are new and unfamiliar with the laboratory environment tend to regard the animals as pets (Arluke 1988, 104).

6. I draw on the argument that "the animal for science is not completely divorced from the animal of common sense," as this seems to apply to producing-food animals, too (Lynch 1988, 282).

7. James Serpell (2004, 150) notes that this affect-utility model is hypothetical and needs to be fully validated; moreover, it "refers specifically to people's attitudes to animals rather than their attitudes to the particular ways in which animals are used, exploited and disposed of."

8. This could apply to bulls and show animals.

9. Recent work has considered additional factors, such as how the animals are housed, generational continuity of livestock bloodlines, and whether animals are home-bred or bought in (Bock et al. 2007).

10. The phrase "the troubled middle" is credited to Strachan Donnelley.

Glossary of Doric Terms

A

aa	all, every
aathing	everything
aboot	about
ach	oh
aff	off
aifter	after
arnae	are not
awa	away
ay	yes
aye	always, still

B

bitty	bit
brither	brother
byre	cowshed

C

canna	cannot
chapped	knocked
coorse	course
cudna	could not

D

dae	do
dee	do, die
deein	doing
deen	done
dinna	don't
dis	does
disna	doesn't
doon	down, depressed, fed up
dour	grim, austere

E

eeky	squeamish
eens	ones

F

faar	where
fae	from
fairming	farming
fit	what, which

G

gaan	going
ging	go
gonna	going to

H

hae	have
hail	whole
heid	head
hid	had
hinna	have not
hiv	have

* Denotes words I translated from the context of the interview.

hivna	have not		**R**	
hoose	house		*roon*	round
			roup	auction
I				
isna	is not		**S**	
ither	other		*saa*	saw
			sae	so
J			*safties*	weak (-minded) person
jist	just		*sharn*	cow dung
			sheet	shoot
K			*slauchter-*	slaughter-
ken	know		*hoose*	house
*killt**	killed		*slitin'**	slitting
kin'	kind, sort		*spad*	spade
			spik	speak, subject of talk or gossip
L			*spikin*	speaking
laisted	lasted		*splut**	split
lang	long		*stannin*	standing
			sudna	shouldn't
M				
ma	me, my		**T**	
mair	more		*tae*	to
mak	make		*tak*	take
			till	to
N			*tup*	ram
nae	no, not			
*naebody**	nobody		**W**	
naething	nothing		*waaking*	walking
niver	never		*wee*	small, a little bit, a little while
noo	now		*weel*	well
nowte	cattle		*wi*	with
			wid	would
O			*widna*	would not
o	of, on		*wifie*	woman, married or not
och	oh		*wis*	was
onybody	anybody		*wisna*	was not
onything	anything		*wudna*	would not
oot	out		*wye*	way
orra	odd, idle, worthless, shabby			
			Y	
			ye	you
P			*yer*	your
petty	pet		*yersel*	yourself
pit	put		*ye've*	you have
*pitting**	putting			

References

Adams, Carol. 2000. *The Sexual Politics of Meat: A Feminist-Vegetarian Critical Theory.* 10th ed. New York: Continuum.

Adams, Carol, and Josephine Donovan, eds. 1995. *Animals and Women: Feminist Theoretical Explorations.* Durham, N.C.: Duke University Press.

Alderson, Lawrence. 1993. "Problems Associated with a Declining Population Size in Endangered Breeds of Livestock." *Ark* (August): 289–292.

———. 1994. *The Chance to Survive.* Northamptonshire, U.K.: Pilkington Press.

Anderson, Kay. 1997. "A Walk on the Wild Side: A Critical Geography of Domestication." *Progress in Human Geography* 21 (4): 463–485.

———. 1998. "Animal Domestication in Geographic Perspective." *Society and Animals* 6 (2): 119–135.

Anderson, Kay, and Susan J. Smith, eds. 2001. "Editorial: Emotional Geographies." *Transactions of the Institute of British Geographers* 26 (1): 7–10.

Anderson, Virginia. 2004. *Creatures of Empire: How Domestic Animals Transformed Early America.* Oxford: Oxford University Press.

Animal Aid. 2006. *Live Exports: The Fate of British Dairy Calves.* Tonbridge: Animal Aid. Available online at http://www.animalaid.org.uk (accessed August 27, 2009).

Animal and Plant Health Inspection Service. 2007. *United States Animal Health Report.* U.S. Department of Agriculture, Agriculture Information Bulletin no. 803. Available online at http://www.aphis.usda.gov (accessed August 26, 2009).

Appleby, Michael. 2006. "Animal Sentience in U.S. Farming." In *Animals, Ethics and Trade: The Challenge of Animal Sentience,* ed. Jacky Turner and Joyce D'Silva, 159–165. London: Earthscan.

Arluke, Arnold. 1988. "Sacrificial Symbolism in Animal Experimentation: Object or Pet?" *Anthrozoös* 2 (2): 98–117.

———. 1991. "Coping with Euthanasia: A Case Study of Shelter Culture." *Journal of the American Veterinary Medical Association* 198 (7): 1176–1180.

———. 1994. "Managing Emotions in an Animal Shelter." In *Animals and Human Society: Changing Perspectives,* ed. Aubrey Manning and James Serpell, 145–165. London: Routledge.

———. 2004. *Brute Force: Animal Police and the Challenge of Cruelty.* West Lafayette, Ind.: Purdue University Press.

———. 2006. *Just a Dog: Understanding Animal Cruelty and Ourselves.* Philadelphia: Temple University Press.

Arluke, Arnold, and Frederic Hafferty. 1996. "From Apprehension to Fascination with 'Dog Lab': The Use of Absolutions by Medical Students." *Journal of Contemporary Ethnography* 25 (2): 201–225.

Arluke, Arnold, and Clinton Sanders. 1996. *Regarding Animals.* Philadelphia: Temple University Press.

Baker, Steve. 1993. *Picturing the Beast: Animals, Identity and Representation.* Manchester, U.K.: Manchester University Press.

Balaam, Nick, Iris Barry, Warwick Bray, et al. 1977. *Illustrated Dictionary of Archaeology.* London: Trewin Copplestone.

Barlett, Peggy. 2006. "Three Visions of Masculine Success on American Farms." In *Country Boys: Masculinity and Rural Life,* ed. Hugh Campbell, Mayerfeld Bell, and Margaret Finney, 47–65. State College: Pennsylvania State University Press.

Benyon, N. M. 1991. "Pig–Primate Interface: Analysis of Stockmanship." *Pig Veterinary Journal* 26:67–77.

Berger, John. 1980. *About Looking.* New York: Pantheon Books.

Berger, Peter, Brigitte Berger, and Kellner Hansfried. 1973. *The Homeless Mind: Modernization and Consciousness.* New York: Random House.

Bertenshaw, Catherine, and Peter Rowlinson. 2009. "Exploring Stock Managers' Perceptions of the Human–Animal Relationship on Dairy Farms and an Association with Milk Production." *Anthrozoös* 22 (1): 59–69.

Birke, Lynda, Arnold Arluke, and Mike Michael. 2007. *The Sacrifice: How Scientific Experiments Transform Animals and People.* West Lafayette, Ind.: Purdue University Press.

Bock, Bettina, M. Van Huik, Madeleine Prutzer, Florence Kling-Eveillard, and Anne-Charlotte Dockès. 2007. "Farmers' Relationship with Different Animals: The Importance of Getting Close to the Animals. Case Studies of French, Swedish and Dutch Cattle, Pig and Poultry Farmers." *International Journal of Sociology of Food and Agriculture* 15 (3): 108–125.

Boeck, George. 1990. *Texas Livestock Auctions: A Folklife Ethnography.* New York: AMS Press.

Boivin, Xavier, Joop Lensink, Céline Tallet, and Isabelle Veissier. 2003. "Stockmanship and Farm Animal Welfare." *Animal Welfare* 12 (4): 479–492.

Bökönyi, Sándor. 1989. "Definitions of Animal Domestication." In *The Walking Larder: Patterns of Domestication, Pastoralism, and Predation,* ed. Juliet Clutton-Brock, 22–27. London: Unwin Hyman.

Boyd, Stephen. 1998. *Hobby Farming—for Pleasure or Profit?* Working Paper no. 33. Ottawa: Statistics Canada Agriculture Division.

Brambell, F. W. Rogers. 1965. *Report of the Technical Committee to Enquire into the Welfare of Animals Kept under Intensive Livestock Husbandry Systems.* London: Her Majesty's Stationery Office.

Brandth, Berit. 1995. "Rural Masculinity in Transition: Gender Images in Tractor Advertisements." *Journal of Rural Studies* 11 (2): 123–133.

———, 2002a. "Gender Identity in European Family Farming. A Literature Review." *Sociologia Ruralis* 42 (3): 181–200.

———. 2002b. "On the Relationship between Feminism and Farm Women." *Agriculture and Human Values* 19 (2): 107–117.

Brayer, Herbert. 1949. "The Influence of British Capital on the Western Range-Cattle Industry." *Journal of Economic History* 9:85–98.

Brown, Jonathan. 2006. "Export of Live Veal Calves to Resume despite Protests." *Independent,* March 6.

Bruinsma, Jelle, ed. 2003. *World Agriculture: Towards 2015/2030, an FAO Perspective.* London: Earthscan.

Bryant, John. 1990. *Fettered Kingdoms: An Examination of a Changing Ethic.* Colden Common near Winchester, Hants, U.K.: Fox Press.

Budiansky, Stephen. 1992. *The Covenant of the Wild: Why Animals Chose Domestication.* New York: William Morrow.

Buller, Henry, and Carol Morris. 2003. "Farm Animal Welfare: A New Repertoire of Nature–Society Relations or Modernism Re-embedded?" *Sociologia Ruralis* 43 (3): 216–237.

Burnell, Ross. 1999. "Hobby Farmers Encouraged to Seek Expert Advice on Animal Husbandry: Ministry of Agriculture and Forestry." Available online at http://www.maf.govt.nz/mafnet/press/archive/1999/110599hob.htm (accessed February 28, 2002).

Burt, Jonathan. 2006. "Conflicts around Slaughter in Modernity." In *Killing Animals,* ed. Animal Studies Group, 120–144. Urbana: University of Illinois Press.

Cameron, David. 1978. *The Ballad and the Plough: A Portrait of the Life of the Old Scottish Farmtouns.* London: Futura Publications.

———. 1995. *The Cornkister Days: A Portrait of a Land and Its Rituals.* Edinburgh: Birlinn.

Camm, Tara, and David Bowles. 2000. "Animal Welfare and the Treaty of Rome—a Legal Analysis of the Protocol on Animal Welfare and Welfare Standards in the European Union." *Journal of Environmental Law* 12 (2): 197–205.

Carlson, Laurie. 2001. *Cattle: An Informal Social History.* Chicago: Ivan R. Dee.

Carlyle, William. 1975. "Livestock Markets in Scotland." *Annals of the Association of American Geographers* 65 (3): 449–460.

Carter, Ian. 1979. *Farm Life in Northeast Scotland 1840–1914: The Poor Man's Country.* Edinburgh: John Donald Publishers.

Cassidy, Ralph. 1967. *Auctions and Auctioneering.* Berkeley: University of California Press.

Cassidy, Rebecca. 2007. "Introduction: Domestication Reconsidered." In *Where the Wild Things Are Now: Domestication Reconsidered,* ed. Rebecca Cassidy and Molly Mullin, 1–25. Oxford: Berg.

Cassidy, Rebecca, and Molly Mullin, eds. 2007. *Where the Wild Things Are Now: Domestication Reconsidered.* Oxford: Berg.

Christie, Les. 2005. "Farming for Fun." January 6. Available online at http://cnn-money.printthis.clickability.com/pt/cpt?action=cpt&title=Hobby+farms (accessed September 20, 2008).

Clutton, O. 1991. "Why We Must Preserve Our Rare Breeds." *Smallholder* (August): 28–29.

Clutton-Brock, Juliet. 1989. "Introduction to Domestication." In *The Walking Larder: Patterns of Domestication, Pastoralism, and Predation,* ed. Juliet Clutton-Brock, 7–9. London: Unwin Hyman.

———. 1994. "The Unnatural World: Behavioural Aspects of Humans and Animals in the Process of Domestication." In *Animals and Human Society: Changing Perspectives,* ed. Aubrey Manning and James Serpell, 23–35. London: Routledge.

———. 1999. *A Natural History of Domesticated Mammals,* 2nd ed. Cambridge: Cambridge University Press.

Cockburn, Alexander. 1996. "A Short, Meat-Orientated History of the World: From Eden to the Mattole." *New Left Review* 215:16–42.

Collins. 1993. *English Dictionary and Thesaurus.* Glasgow: HarperCollins.

Commission for Rural Communities. 2007. *The State of the Countryside 2007.* Gloucestershire, U.K.: Commission for Rural Communities.

Convery, Ian, Cathy Bailey, Maggie Mort, and Josephine Baxter. 2005. "Death in the Wrong Place? Emotional Geographies of the U.K. 2001 Foot and Mouth Disease Epidemic." *Journal of Rural Studies* 21 (1): 99–109.

Cronon, William. 1991. *Nature's Metropolis: Chicago and the Great West.* New York: W. W. Norton.

Dalzell, Ian. 2000. "Hobby Farming Haven." Available online at http://www.nfucountry-side.org.uk/news/jan00/jan16.htm (accessed March 13, 2002).

Daniels, Thomas. 1986. "Hobby Farming in America: Rural Development or Threat to Commercial Agriculture?" *Journal of Rural Studies* 2 (1): 31–40.

Davis, David. 1984. "Good People Doing Dirty Work: A Study of Social Isolation." *Symbolic Interaction* 7 (2): 233–247.

DeGrazia, David. 2003. "Meat-Eating." In *The Animal Ethics Reader,* ed. Susan Armstrong and Richard Botzler, 177–183. London: Routledge.

Delgado, Christopher, Mark Rosegrant, Henning Steinfeld, Simeon Ehui, and Claude Courbois. 1999. *Livestock to 2020: The Next Food Revolution.* Food Agriculture, and the Environment Discussion Paper 28. Washington, D.C.: International Food Policy Research Institute, Food and Agriculture Organization of the United Nations, and International Livestock Research Institute.

Department for Environment, Food, and Rural Affairs. 2007a. "Clinical Signs of Bovine Spongiform Encephalopathy in Cattle." Transmissible Spongiform Encephalopathy European Community Reference Laboratory. Available online at http://www.defra. gov.uk (accessed August 22, 2009).

———. 2007b. "Livestock Movements, Identification and Tracing: Cattle—Information for EU Member States." Available online at http://www.defra.gov.uk (accessed March 11, 2009).

———. 2008. "FMD: Comparisons with the 1967 FMD Outbreak." Available online at http://www.defra.gov.uk (accessed March 11, 2009).

———. 2009. "What Is Electronic Identification (EID)?" Available online at http://www. defra.gov.uk (accessed December 21, 2009).

Digard, Jean-Pierre. 1994. "Relationships between Humans and Domesticated Animals." *Interdisciplinary Science Reviews* 19 (3): 231–236.

Dimich-Ward, H., J. R. Guernsey, W. Pickett, D. Rennie, L. Hartling, and R. J. Brison. 2004. "Gender Differences in the Occurrence of Farm Related Injuries." *Occupational and Environmental Medicine* 61 (1): 52–56.

Dixon, Jane. 2002. *The Changing Chicken: Chooks, Cooks and Culinary Culture.* Sydney: University of New South Wales Press.

Dockès, Anne Charlotte, and Florence Kling-Eveillard. 2006. "Farmers' and Advisers' Representations of Animals and Animal Welfare." *Livestock Science* 103 (3): 243–249.

Dowling, Robert, Lawrence Alderson, and Roger Caras. 1994. *Rare Breeds.* London: Lawrence King.

Druce, Clare, and Philip Lymbery. 2006. "Outlawed in Europe." In *In Defense of Animals: The Second Wave,* ed. Peter Singer, 123–131. Malden, Mass.: Blackwell.

Ducos, Pierre. 1989. "Defining Domestication: A Clarification." In *The Walking Larder: Patterns of Domestication, Pastoralism, and Predation,* ed. Juliet Clutton-Brock, 28–30. London: Unwin Hyman.

Dunayer, Joan. 2001. *Animal Equality: Language and Liberation.* Derwood, Md.: Ryce Publishing.

Duncan, Ian. 2004. "A Concept of Welfare Based on Feelings." In *The Well-being of Farm Animals: Challenges and Solutions,* ed. John Benson and Bernard Rollin, 85–101. Ames, Iowa: Blackwell.

———. 2006. "The Changing Concept of Animal Sentience." *Applied Animal Behaviour Science* 100 (1–2): 11–19.

DuPuis, E. Melanie. 2002. *Nature's Perfect Food: How Milk Became America's Drink.* New York: New York University Press.

Eisnitz, Gail. 1997. *Slaughterhouse: The Shocking Story of Greed, Neglect, and Inhumane Treatment inside the U.S. Meat Industry.* New York: Prometheus Books.

Elliot, Valerie. 2008. "Smallholders Hit Back at Claim that They Threaten Future of British Farming." *Times,* February 18.

Ellis, Colter. 2007. "Negotiating Contradiction: Human–Animal Relationships in Cattle Ranching." Paper presented at the annual meeting of the American Sociological Association, New York, October 22.

Ellis, Colter, and Leslie Irvine. 2008. "Reproducing Dominion: Emotional Apprenticeship in the 4H Youth Livestock Program." Paper presented at the annual meeting of the American Sociological Association, Boston, July 31.

Ellis, Hattie. 2007a. *Planet Chicken: The Shameful Story of the Bird on Your Plate.* London: Hodder and Stoughton.

Emel, Jody, and Jennifer Wolch. 1998. "Witnessing the Animal Moment." In *Animal Geographies: Place, Politics and Identity in the Nature–Culture Borderlands,* ed. Jennifer Wolch and Jody Emel, 1–24. London: Verso.

English, Peter. 2002. "Overview of the Evaluation of Stockmanship." Paper presented at the National Pork Board Symposium on Swine Housing and Well-being, Des Moines, Iowa, June 5, pp. 1–8.

English, Peter, Gethyn Burgess, Ricardo Segundo, and John Dunne. 1992. *Stockmanship: Improving the Care of the Pig and Other Livestock.* Ipswich, U.K.: Farming Press.

English, Peter, and Owen McPherson. 1998a. "Improving Stockmanship in Pig Production and the Role of the EU Leonardo Initiatives." Available online at www.gov.on.ca/OMAFRA/english/livestock/swine/facts/article1.htm (accessed April 8, 2002).

———. 1998b. "Improving the Quality of Pig Stockmanship by Attracting Better Job Applicants and by Selecting and Retaining the Best." Available online at www.gov.on.ca/OMAFRA/english/livestock/swine/facts/article5.htm (accessed April 8, 2002).

European Commission. 2002. *Farm Animal Welfare: Current Research and Future Directions.* Luxembourg: Office for Official Publications of the European Communities.

Evans, Nick, and Brian Ilbery. 1993. "The Pluriactivity, Part-Time Farming, and Farm Diversification Debate." *Environment and Planning A* 25 (7): 945–959.

Evans, Nick, Carol Morris, and Michael Winter. 2002. "Conceptualizing Agriculture: A Critique of Post-productivism as the New Orthodoxy." *Progress in Human Geography* 26 (3): 313–332.

Evans, Nick, and Richard Yarwood. 1995. "Livestock and Landscape." *Landscape Research* 20 (3): 141–146.

———. 1998. "Results of the Membership Survey." *Ark* (Spring): 10, 16.

———. 2000. "The Politicization of Livestock: Rare Breeds and Countryside Conservation." *Sociologia Ruralis* 40 (2): 228–248.

Evans-Pritchard, Edward E. 1940. *The Nuer: A Description of the Modes of Livelihood and Political Institutions of Nilotic People.* Oxford: Clarendon Press.

Farm Animal Welfare Council. 2007. *FAWC Report on Stockmanship and Farm Animal Welfare.* London: Farm Animal Welfare Council.

Fiddes, Nick. 1991. *Meat: A Natural Symbol.* London: Routledge.

Fink, Deborah. 1986. *Open Country, Iowa: Rural Women, Tradition and Change.* Albany: State University of New York Press.

Food and Agriculture Organization. 1995. *Women, Agriculture and Rural Development: A Synthesis Report of the North East Region—Women, Agriculture and Rural Development.* Rome: Food and Agriculture Organization.

———. 2000. *Livestock and Gender: A Winning Pair.* Working document by the Swiss Agency for Development and Cooperation. Bern. Available online at http://www.fao.org (accessed August 2, 2007).

———. 2006. *Livestock's Long Shadow: Environmental Issues and Options.* Rome: Food and Agriculture Organization.

Food Standards Agency. 2001. "Restriction on Pithing (England) Regulations 2001." Available online at http://www.food.gov.uk (accessed March 16, 2009).

———. 2008. "FSA Board Advise on BSE Testing Age." Available online at http://www.food.goc.uk (accessed March 17, 2009).

Fox, Michael. 1999. "American Agriculture's Ethical Crossroads." In *The Meat Business: Devouring a Hungry Planet,* ed. Geoff Tansey and Joyce D'Silva, 25–42. London: Earthscan.

France, Louise. 2008. "Meet the Greenshifters." *Observer Magazine,* September 7, pp. 18–26.

Francione, Gary. 1998. "Animal Rights and New Welfarism." In *Encyclopedia of Animal Rights and Animal Welfare,* ed. Marc. Bekoff, 45. Westport, Conn.: Greenwood Press.

———. 2006. "Animals, Property, and Personhood." In *People Property or Pets?* ed. Marc Hauser, Fiery Cushman, and Matthew Kamen, 77–102. West Lafayette, Ind.: Purdue University Press.

Franklin, Adrian. 1999. *Animals and Modern Cultures: A Sociology of Human–Animal Relations in Modernity.* London: Sage.

Franklin, Sarah. 2007. *Dolly Mixtures: The Remaking of Genealogy.* Durham, N.C.: Duke University Press.

Fraser, Allan. 1960. *Animal Husbandry Heresies.* London: Crosby Lockwood and Son.

Fraser, David. 2001. "The 'New Perception' of Animal Agriculture: Legless Cows, Featherless Chickens, and a Need for Genuine Analysis." *Journal of Animal Science* 79 (3): 634–641.

Fukuda, Kaoru. 1996. "The Place of Animals in British Moral Discourse: A Field Study from the Scottish Borders." Ph.D. diss., Oxford University.

Gaard, Greta, ed. 1993. *Ecofeminism: Women, Animals, Nature.* Philadelphia: Temple University Press.

Garner, Frank. 1946. *The Farmer's Animals: How They Are Bred and Reared.* Cambridge: Cambridge University Press.

Garner, Robert. 1998. "The Politics of Farm Animal Welfare in Britain." In *Political Animals: Animal Protection Politics in Britain and the United States,* ed. Robert Garner, 151–175. Hampshire, U.K.: Macmillan Press.

Gasson, Ruth. 1980. "Roles of Farm Women in England." *Sociologia Ruralis* 20 (3): 165–180.

———. 1988. *The Economics of Part-Time Farming.* Essex, U.K.: Longman Scientific and Technical.

———. 1990. "Part-Time Farming and Pluriactivity." In *Agriculture in Britain: Changing Pressures and Policies,* ed. Denis Britton, 161–172. Oxon, U.K.: CAB International.

Gatward, Gordon. 2001. *Livestock Ethics: Respect, and Our Duty of Care for Farm Animals.* Lincoln, U.K.: Chalcombe Publications.

Gellatley, Juliet. 1996. *The Silent Ark: A Chilling Exposé of Meat—the Global Killer.* London: Thorsons.

Glenn, Cathy B. 2004. "Constructing Consumables and Consent: A Critical Analysis of Factory Farm Industry Discourse." *Journal of Communication Inquiry* 28 (63): 63–81.

Goffman, Erving. 1963. *Stigma: Notes on the Management of Spoiled Identity.* London: Penguin Books.

Graham, Ian. 1999. "The Construction of Electronic Markets." Ph.D. diss., Edinburgh University.

Grandin, Temple. 2000. "Introduction: Management and Economic Factors of Handling and Transport." In *Livestock Handling and Transport,* 2nd ed., ed. Temple Grandin, 114. Oxon, U.K.: Commonwealth Agricultural Bureaux International Publishing.

———. 2006a. "Animals Are Not Things." In *People Property or Pets?* ed. Marc Hauser, Fiery Cushman, and Matthew Kamen, 205–211. West Lafayette, Ind.: Purdue University Press.

———. 2006b. "Progress and Challenges in Animal Handling and Slaughter in the U.S." *Applied Animal Behaviour Science* 100 (1–2): 129–139.

Grasseni, Cristina. 2004. "Skilled Vision: An Apprenticeship in Breeding Aesthetics." *Social Anthropology* 12 (1): 41–55.

―――. 2005. "Designer Cows: The Practice of Cattle Breeding Between Skill and Standardization." *Society and Animals* 13 (1): 33–49.

Gray, John. 1998. "Family Farms in the Scottish Borders: A Practical Definition by Hill Sheep Farmers." *Journal of Rural Studies* 14 (3): 341–356.

Great Britain. 1822. *An Act to Prevent the Cruel Improper Treatment of Cattle* (Martin's Act). London: George Eyre and Andrew Strachan.

Gregory, Neville. 1984. "Animal Welfare in the Hauling and Slaughtering Trades: Anecdotes from the Past." *Biologist* 31 (1): 23–26.

―――. 1989–1990. "Slaughtering Methods and Equipment." *Veterinary History New Series* 6 (2): 73–84.

Gurian-Sherman, Doug. 2008. *CAFOs Uncovered: The Untold Costs of Confined Animal Feeding Operations.* Cambridge, Mass.: Union of Concerned Scientists.

Haldane, A. 2002. *The Drove Roads of Scotland.* Edinburgh: Birlinn.

Halfacree, Keith. 2007. "Back-to-the-Land in the Twenty-first Century—Making Connections with Rurality." *Tijdschrift voor Economische en Sociale Geografie* 98 (1): 3–8.

Hammersley, Martin, and Paul Atkinson. 1995. *Ethnography: Principles in Practice,* 2nd ed. London: Routledge.

Hardy, Anne. 2003. "Professional Advantage and Public Health: British Veterinarians and State Veterinary Services, 1865–1939." *Twentieth Century British History* 14 (1): 1–23.

Harrington, Lisa, and Max Lu. 2002. "Beef Feedlots in Southwestern Kansas: Local Change, Perceptions, and the Global Change Context." *Global Environmental Change* 12:273–282.

Harrison, Ruth. 1964. *Animal Machines.* London: Stuart.

―――. 1970. "Steps towards Legislation in Great Britain." In *Factory Farming: A Symposium,* ed. J. Bellerby, 3–16. Oxford: Alden Press.

Harvey, Matthew. 2007. "Animal Genomics in Science, Social Science and Culture." *Genomics, Society and Policy* 3 (2): 1–28.

Haughen, Marit, and Berit Brandth. 1994. "Gender Differences in Modern Agriculture: The Case of Female Farmers in Norway." *Gender and Society* 8 (2): 206–229.

Health and Safety Executive. 2006. *Exploring the Influences of Farming Women and Families on Worker Health and Safety.* Research Report no. 423. Norwich, U.K.: Her Majesty's Stationery Office.

Hecker, Howard. 1982. "Domestication Revisited: Its Implications for Faunal Analysis." *Journal of Field Archaeology* 9 (2): 217–236.

Hemsworth, Paul. 2000. "Stockmanship Makes a Difference." Available online at http://mark.asci.edu/HealthyHogs/book2000/hemsworth.htm (accessed April 8, 2002).

―――. 2004. "Human–Livestock Interaction." In *The Well-being of Farm Animals: Challenges and Solutions,* ed. John Benson and Bernard Rollin, 21–38. Oxford: Blackwell.

―――. 2007. "Ethical Stockmanship." *Australian Veterinary Journal* 85 (5): 194–200.

Hemsworth, Paul, and Grahame Coleman. 1998. *Human–Livestock Interactions: The Stockperson and the Productivity and Welfare of Intensively Farmed Animals.* Oxon, U.K.: Commonwealth Agricultural Bureaux International.

Heritage, John. 1984. *Garfinkel and Ethnomethodology.* Cambridge: Polity Press.

Herzog, Harold. 1993. "Human Morality and Animal Research: Confessions and Quandaries." *American Scholar* 62 (3): 337–349.

Herzog, Harold, Nancy Betchart, and Robert Pittman. 1991. "Gender, Sex Role Orientation, and Attitudes toward Animals." *Anthrozoös* 4 (3): 184–191.

Herzog, Harold, and S. McGee. 1983. "Psychological Aspects of Slaughter: Reactions of College Students to Killing and Butchering Cattle and Hogs." *International Journal Study of Animal Problems* 4 (2): 124–132.

Hochschild, Arlie Russell. 1979. "Emotion Work, Feeling Rules, and Social Structure." *American Journal of Sociology* 85 (3): 551–575.

———. 1983. *The Managed Heart*. Berkeley: University of California Press.

Hodge, Ian. 1990. "The Changing Place of Farming." In *Agriculture in Britain: Changing Pressures and Policies*, ed. Denis Britton, 34–44. Oxon, U.K.: Commonwealth Agricultural Bureaux International.

Holloway, Lewis. 2000a. "'Hell on Earth and Paradise All at the Same Time': The Production of Smallholding Space in the British Countryside." *Area* 32 (3): 307–315.

———. 2000b. "Hobby-Farming in the U.K.: Producing Pleasure in the Post-productivist Countryside." II Anglo Spanish Symposium on Rural Geography, University of Valladolid, Spain, July 1–11.

———. 2001. "Pets and Protein: Placing Domestic Livestock on Hobby-Farms in England and Wales." *Journal of Rural Studies* 17 (3): 293–307.

———. 2002. "Smallholding, Hobby-Farming, and Commercial Farming: Ethical Identities and the Production of Farming Spaces." *Environment and Planning A* 34 (11): 2055–2070.

———. 2005. "Aesthetics, Genetics, and Evaluating Animal Bodies: Locating and Displacing Cattle on Show and in Figures." *Environment and Planning D: Society and Space* 23 (6): 883–902.

———. 2007. "Subjecting Cows to Robots: Farming Technologies and the Making of Animal Subjects." *Environment and Planning D: Society and Space* 25 (6): 1041–1060

Hoppe, Robert, ed. 2001. *Structural and Financial Characteristics of U.S. Farms: 2001 Family Farm Report*. Resource Economics Division, Economic Research Service, U.S. Department of Agriculture. Agriculture Information Bulletin no. 768.

Horowitz, Roger. 2006. *Putting Meat on the American Table: Taste, Technology, Transformation*. Baltimore: Johns Hopkins University Press.

Howarth, Richard. 2000. "The CAP: History and Attempts at Reform." *Economic Affairs* 29 (2): 4–10.

HRH The Prince of Wales. 1998. "Letter in Comment Page." *Ark* (Fall): 93.

Huey, Robert. 1998. "Meat Hygiene: Is It a Job for a Vet?" *Continuing Professional Development Veterinary Medicine* 1 (2): 43–45.

Humphrey, Nicholas. 1995. "Histories." *Social Research* 62 (3): 477–479.

Hunter, Kathryn, and Pamela Riney-Kehrberg. 2002. "Rural Daughters in Australia, New Zealand and the United States: An Historical Perspective." *Journal of Rural Studies* 18 (2): 135–143.

Ilbery, Brian. 1988. "Farm Diversification and the Restructuring of Agriculture." *Outlook on Agriculture* 17 (1): 35–39.

———. 1991. "Farm Diversification as an Adjustment Strategy on the Urban Fringe of the West Midlands." *Journal of Rural Studies* 7 (3): 207–218.

Ingold, Tim. 1994. "From Trust to Domination: An Alternative History of Human–Animal Relations." In *Animals and Human Society: Changing Perspectives,* ed. Aubrey Manning and James Serpell, 1–22. London: Routledge.

———. 2000. *The Perception of the Environment: Essays in Livelihood, Dwelling and Skill.* London: Routledge.

International Livestock Research Institute. 2007. *Safeguarding Livestock Diversity: The Time Is Now.* Nairobi: Regal Press.

Irvine, Leslie. 2003. "The Problem of Unwanted Pets: A Case Study in How Institutions 'Think' about Clients' Needs." *Social Problems* 50 (4): 550–566.

———. 2004. *If You Tame Me: Understanding Our Connection with Animals.* Philadelphia: Temple University Press.

Jackson, G. 1998. "A Place in the Market for the Rare Breeds." *Farmer's Club Journal,* no. 156 (Fall): 10–11.

Jamison, Wesley, and William Lunch. 1992. "Rights of Animals, Perceptions of Science, and Political Activism: Profile of American Animal Rights Activists." *Science, Technology, and Human Values* 17 (4): 438–458.

Janowitz, Morris, ed. 1966. "The Relation of Research to the Social Process." In *W. I. Thomas on Social Organization and Social Personality,* 289–305. Chicago: University of Chicago Press.

Jedrej, Charles, and Mark Nuttall. 1996. *White Settlers: The Impact of Rural Repopulation in Scotland.* Luxembourg: Harwood Academic Publishers.

Kahrs, Robert. 2004. *Global Livestock Health Policy: Challenges, Opportunities, and Strategies for Effective Action.* Ames: Iowa State Press.

Kendall, Holli, Linda Lobao, and Jeff Sharp. 2006. "Public Concern with Well-being: Place, Social Structural Location, and Individual Experience." *Rural Sociology* 71 (3): 399–428.

Kenny, Keith. 2006. "McDonald's: Progressing Global Standards in Animal Welfare." In *Animals, Ethics and Trade: The Challenge of Animal Sentience,* ed. Jacky Turner and Joyce D'Silva, 166–174. London: Earthscan.

Kim, Chul-Kyoo, and James Curry. 1993. "Fordism, Flexible Specialization and Agri-industrial Restructuring: The Case of the U.S. Broiler Industry." *Sociologia Ruralis* 33 (1): 61–80.

King George V (23 and 24). 1933. *Slaughter of Animals Act.* London: Eyre and Spottiswoode.

Knapp, Vincent. 1997. "The Democratization of Meat and Protein in Late Eighteenth- and Nineteenth-Century Europe." *Historian* 59 (3): 541–551.

Kopytoff, Igor. 1986. "The Cultural Biography of Things: Commoditization as Process." In *The Social Life of Things: Commodities in Cultural Perspective,* ed. Arjun Appadurai, 64–91. Cambridge: Cambridge University Press.

Kouimintzis, Dimitris, Christos Chatzis, and Athena Linos. 2007. "Health Effects of Livestock Farming in Europe." *Journal of Public Health* 15 (4): 245–254.

Kynoch, Douglas. 1996. *A Doric Dictionary: Two-Way Lexicon of North-East Scots (Doric~English/English~Doric).* Edinburgh: Scottish Cultural Press.

Lang, Tim, and Michael Heasman. 2004. *Food Wars: The Global Battle for Mouths, Minds and Markets.* London: Earthscan.

Lantra. 2009. *Women and Work: Raising Skills and Unlocking Potential.* Available online at http://www.lantra.co.uk (accessed August 13, 2009).

Lawrence, Elizabeth. 1989. "Neoteny in American Perceptions of Animals." In *Perceptions of Animals in American Culture,* ed. R. J. Hoage, 57–76. Washington, D.C.: Smithsonian Institution Press.

Lawrence, Felicity. 2004. *Not on the Label: What Really Goes into the Food on Your Plate.* London: Penguin Books.

Layton, R. 1978. "The Operational Structure of the Hobby Farm." *Area* 10 (4): 242–246.

———. 1980. "Hobby Farming." *Geography* 65 (1): 220–224.

Leach. Helen. 2007. "Selection and the Unforeseen Consequences of Domestication." In *Where the Wild Things Are Now: Domestication Reconsidered,* ed. Rebecca Cassidy and Molly Mullin, 71–99. Oxford: Berg.

Leckie, Gloria. 1996a. "Female Farmers and the Social Construction of Access to Agricultural Information." *Library and Information Science* 18 (4): 297–321.

———. 1996b. "'They Never Trusted Me to Drive': Farm Girls and the Gender Relations of Agricultural Information Transfer." *Gender, Place and Culture* 3 (3): 309–325.

Lensink, Joop, Alain Boissy, and Isabelle Veissier. 2000. "The Relationship between Farmer's Attitude and Behaviour towards Calves, and Productivity of Veal Units." *Annales de Zootechnie* 49 (4): 313–327.

Levinger, I. 1979. "Jewish Attitude toward Slaughter." *Animal Regulation Studies* 2:103–109.

Ley, John. 1978. *The Story of My Life: Recollections of a North Devon Farmer, Preacher, and Family Man.* Elms Court, U.K.: Arthur H. Stockwell.

Liepins, Ruth. 2000. "Making Men: The Construction and Representation of Agriculture-Based Masculinities in Australia and New Zealand." *Rural Sociology* 65 (4): 605–620.

Lindsay, Sandra, Sivasubramaniam Selvaraj, John W. Macdonald, and David J. Godden. 2004. "Injuries to Scottish Farmers while Tagging and Clipping Cattle: A Cross-Sectional Survey." *Occupational Medicine* 54 (2): 86–91.

Little, Jo. 2002. "Rural Geography: Rural Gender Identity and the Performance of Masculinity and Femininity in the Countryside." *Progress in Human Geography* 26 (5): 665–670.

Livestock Knowledge Transfer. 2001. "Carcass Classification." Fact Sheet: Beef and Sheep 314.

Lofstedt, Jeanne. 2003. "Gender and Veterinary Medicine." *Canadian Veterinary Journal* 44:533–535.

Lovenheim, Peter. 2002. *Portrait of a Burger as a Young Calf: The Story of One Man, Two Cows, and the Feeding of a Nation.* New York: Three Rivers Press.

Lulka, David. 2006. "Grass or Grain? Assessing the Nature of the U.S. Bison Industry." *Sociologia Ruralis* 46 (3): 173–191.

———. 2008. "The Paradoxical Nature of Growth in the U.S. Bison Industry." *Journal of Cultural Geography* 25 (1): 31–56.

Lutwyche, Richard. 1998a. "Eating Rare Breeds." In *Smallholder: For Practical Smallfarming Today.* Rare Breeds Survival Trust Special (September), ix.

———. 1998b. "What Is a Rare Breed . . . and Why Is It Rare?" In *Smallholder: For Practical Smallfarming Today.* Rare Breeds Survival Trust Special (September), i–ii.

Lymbery, Philip. 1999. "Campaigning for Change in the European Union." In *The Meat Business: Devouring a Hungry Planet*, ed. Geoff Tansey and Joyce D'Silva, 73–81. London: Earthscan.

Lynch. Michael. 1988. "Sacrifice and the Transformation of the Animal Body into a Scientific Object: Laboratory Culture and Ritual Practice in the Neurosciences." *Social Studies of Science* 18 (2): 265–289.

Maehle, Andreas-Holger. 1994. "Cruelty and Kindness to the 'Brute Creation': Stability and Change in the Ethics of the Human–Animal Relationship, 1600–1850." In *Animals and Human Society: Changing Perspectives*, ed. Aubrey Manning and James Serpell, 81–105. London: Routledge.

Maret, Elizabeth. 1993. *Women of the Range: Women's Roles in the Texas Beef Cattle Industry*. College Station: Texas A&M University Press.

Mason, Jim, and Mary Finelli. 2006. "Brave New Farm?" In *In Defense of Animals: The Second Wave*, ed. Peter Singer, 104–122. Malden, Mass.: Blackwell.

Mason, Jim, and Peter Singer. 1980. *Animal Factories*. New York: Crown Publishers.

Mather, Alexander, Gary Hill, and Maria Nijnik. 2006. "Post-productivism and Rural Land Use: Cul de Sac or Challenge for Theorization?" *Journal of Rural Studies* 22 (4): 441–455.

May, Tim. 2001. *Social Research: Issues, Methods and Process*, 3rd ed. Buckingham: Open University Press.

McLeod, Rhoda. 1998. "Calf Exports at Brightlingsea." In *Protest Politics: Cause Groups and Campaigns*, ed. F. F. Ridley and Grant Jordan, 37–49. New York: Oxford University Press.

M'Combie, William. 1875. *Cattle and Cattle-Breeder*, 3rd ed. Edinburgh: William Blackwood and Sons.

Meadow, Richard. 1989. "Osteological Evidence for the Process of Animal Domestication." In *The Walking Larder: Patterns of Domestication, Pastoralism, and Predation*, ed. Juliet Clutton-Brock, 80–90. London: Unwin Hyman.

Meara, Hannah. 1974. "Honor in Dirty Work: The Case of American Meat Cutters and Turkish Butchers." *Work and Occupations* 1 (3): 259–283.

Meat and Livestock Commission. 2008. *Supermarket Meat Retailing: A Special Report Prepared by the Meat and Livestock Commission on Behalf of the National Farmers' Union*. Milton Keynes, U.K.: Meat and Livestock Commission.

Merton, Robert. 1976. *Sociological Ambivalence and Other Essays*. New York: Free Press.

Mills. C. Wright. 1940. "Situated Actions and the Vocabularies of Motive." *American Sociological Review* 5 (6): 904–913.

Mithen, Steven. 1996. *The Prehistory of the Mind: A Search for the Origins of Art, Religion and Science*. London: Phoenix.

Morris, Carol, and Nick Evans. 2001. "'Cheese Makers Are Always Women': Gendered Representations of Farm Life in the Agricultural Press." *Gender, Place and Culture* 8 (4): 375–390.

National Farmers Union. 2008a. "Sheep EID Bulletin Special." Warwickshire, U.K.: National Farmers Union.

———. 2008b. "Statement on Smallholders." Available online at www.nfuonline.com (accessed September 16, 2008).

Nerlich, Brigitte. 2004. "Risk, Blame and Culture: Foot and Mouth Disease and the Debate about Cheap Food." In *The Politics of Food,* ed. Marianne Lien and Brigitte Nerlich, 39–57. Oxford: Berg.

Neth, Mary. 1995. *Preserving the Family Farm: Women, Community, and the Foundations of Agribusiness in the Midwest, 1900–1940.* Baltimore: Johns Hopkins University Press.

NewLandOwner. 2008. "Who We Are and Courses." Available online at http://www.newlandowner.co.uk (accessed September 16, 2008).

Nibert, David. 2002. *Animal Rights/Human Rights: Entanglements of Oppression and Liberation.* Lanham, Md.: Rowman and Littlefield.

Nierenberg, Danielle. 2005. *Happier Meals: Rethinking the Global Meat Industry.* World Watch Paper no. 171. Washington, D.C.: Worldwatch Institute.

Noske, Barbara. 1989. *Humans and Other Animals: Beyond the Boundaries of Anthropology.* London: Pluto Press.

Ortner, Sherry. 1974. "Is Female to Male as Nature Is to Culture?" In *Woman, Culture and Society,* ed. Michelle Rosaldo and Louise Lamphere, 67–87. Palo Alto, Calif.: Stanford University Press.

Osterud, Nancy. 1993. "Gender and the Transition to Capitalism in Rural America." *Agricultural History* 67 (2): 14–29

Palmer, Clare. 1997. "The Idea of the Domesticated Animal Contract." *Environmental Values* 6 (4): 411–425.

Pattison, John. 1998. "The Emergence of Bovine Spongiform Encephalopathy and Related Diseases." *Emerging Infectious Diseases* 4 (3): 390–394.

Paul, Elizabeth S., and Anthony L. Podberscek. 2000. "Veterinary Education and Students' Attitudes towards Animal Welfare." *Veterinary Record* 146 (10): 269–272.

Pearson, Geoffrey. 1993. "Talking a Good Fight: Authenticity and Distance in the Ethnographer's Craft." In *Interpreting the Field: Accounts of Ethnography,* ed. Dick Hobbs and Tim May, vii–xx. Oxford: Clarendon Press.

Pedersen, Kirsten Bransholm, and Bente Kjærgård. 2004. "Do We Have Room for Shining Eyes and Cows as Comrades? Gender Perspectives on Organic Farming in Denmark." *Sociologia Ruralis* 44 (4): 373–394.

Penman, Danny. 1996. *The Price of Meat.* London: Victor Gollancz.

Perren, Richard. 1971. "The North American Beef and Cattle Trade with Great Britain, 1870–1914." *Economic History Review New Series* 24 (3): 430–444.

———. 1978. *The Meat Trade in Britain 1840–1914.* London: Routledge and Kegan Paul.

———. 1985. "The Retail and Wholesale Meat Trade 1880–1939." In *Diet and Health in Modern Britain,* ed. Derek Oddy and Derek Miller, 46–80. London: Croom Helm.

Peter, Gregory, Michael Mayerfeld Bell, and Susan Jarnagin. 2000. "Coming Back across the Fence: Masculinity and the Transition to Sustainable Agriculture." *Rural Sociology* 65 (2): 215–233.

Petherick, J. Carol. 2005. "Animal Welfare Issues Associated with Extensive Livestock Production: The Northern Australian Beef Cattle Industry." *Applied Animal Behaviour Science* 92 (3): 211–234.

Pew Commission on Industrial Farm Animal Production. 2008. *Putting Meat on the Table: Industrial Farm Animal Production in America.* Final report. Available online at http://www.ncifap.org (accessed July 8, 2009).

Philo, Chris. 1995. "Animals, Geography, and the City: Notes on Inclusions and Exclusions." *Environmental and Planning D: Society and Space* 13 (6): 655–681.

Pick, Daniel. 1993. *War Machine: The Rationalisation of Slaughter in the Modern Age.* New Haven, Conn.: Yale University Press.

Pilgeram, Ryanne. 2007. "'Ass-Kicking' Women: Doing and Undoing Gender in a U.S. Livestock Auction." *Gender, Work and Organization* 14 (6): 572–595.

Pinchbeck, Ivy. 1969. *Women Workers and the Industrial Revolution, 1750–1850.* London: Frank Cass.

Policy Commission on the Future of Farming and Food. 2002. *Farming and Food: A Sustainable Future.* London: Cabinet Office.

Pollan, Michael. 2006. *The Omnivore's Dilemma: The Search for a Perfect Meal in a Fast-Food World.* London: Bloomsbury.

Porcher, Jocelyne. 2006. "Well-being and Suffering in Livestock Farming: Living Conditions at Work for People and Animals." *Sociologie du Travail* 48 (supp. 1): 56–70.

Press and Journal. 1998. "Calf Cull Scheme May Be Kept On." *Press and Journal,* November 7, 23.

Price, Linda, and Nick Evans. 2006. "From 'As Good as Gold' to 'Gold Diggers': Farming Women and the Survival of British Family Farming." *Sociologia Ruralis* 46 (4): 280–298.

Pringle, Rosemary, and Susan Collings. 1993. "Women and Butchery: Some Cultural Taboos." *Australian Feminist Studies* 17:29–45.

Prothero, Rowland. 1912. *English Farming: Past and Present.* London: Longmans Green.

Pye-Smith, Charlie, and Richard North. 1984. *Working the Land.* London: Temple Smith.

Quality Meat Scotland. 2004. *Planned Carcase Production for the Scottish Sheep Industry.* Trade document, Newbridge, U.K.

Quinn, Michael. 1993. "Corpulent Cattle and Milk Machines: Nature, Art and the Ideal Type." *Society and Animals* 1 (2): 145–157.

Radford. Mike. 2001. *Animal Welfare Law in Britain: Regulation and Responsibility.* Oxford: Oxford University Press.

Rare Breeds Survival Trust. 1996. "Native Breeds—New Rare Breeds Classification." *Ark* (Summer): 55–56.

———. 2008a. "Approved Farm Parks." Available online at www.rbst.org.uk (accessed September 15, 2008).

———. 2008b. "Meat from Rare Breeds." Available online at www.rbst.org.uk (accessed September 20, 2008).

———. 2008c. "Watchlist." Available online at www.rbst.org.uk (accessed September 15, 2008).

———. N.d. "Rare Breeds." Fact sheet for the Food and Farming Education Service, Teachers' Resource Pack. Warwickshire, U.K.: Rare Breeds Survival Trust.

Red Meat Industry Forum. 2007. "Introduction to Beef Production in U.K." Available online at http://www.redmeatindustryforum.org.uk (accessed August 24, 2009).

Regan, Tom. 2004. *The Case for Animal Rights,* rev. ed. Berkeley: University of California Press.

Richards, Sam. 2008. "River Cottage Spring." *Telegraph*, May 24.

Rifkin, Jeremy. 1992. *Beyond Beef: The Rise and Fall of the Cattle Culture*. New York: Plume.

Ritvo, Harriet. 1987. *The Animal Estate: The English and Other Creatures in the Victorian Age*. Cambridge, Mass.: Harvard University Press.

Robinson, D., and H. Hooker. 2006. "The U.K. Veterinary Profession in 2006: The Findings of a Survey of the Profession Conducted by the Royal College of Veterinary Surgeons." Institute for Employment Studies, 1–7. Available online at http://www.employment-studies.co.uk (accessed July 22, 2007).

Rollin, Bernard. 1995. *Farm Animal Welfare: Social Bioethical, and Research Issues*. Ames: Iowa State University Press.

———. 2004. "The Ethical Imperative to Control Pain and Suffering in Farm Animals." In *The Well-being of Farm Animals: Challenges and Solutions*, ed. John Benson and Bernard Rollin, 3–19. Oxford: Blackwell.

Rosenfeld, Rachel. 1985. *Farm Women: Work, Farm, and Family in the United States*. Chapel Hill: University of North Carolina Press.

Ross, Peggy. 1985. "A Commentary on Research on American Farmwomen." *Agriculture and Human Values* 2 (1): 19–30.

Royal Society for Prevention of Cruelty to Animals. 1999. "Freedom Food." Information leaflet.

Russell, Nerissa. 2007. "The Domestication of Anthropology." In *Where the Wild Things Are Now: Domestication Reconsidered*, ed. Rebecca Cassidy and Molly Mullin, 27–48. Oxford: Berg.

Ryder, Richard. 2000. *Animal Revolution: Changing Attitudes towards Speciesism*. Oxford: Berg.

Sachs, Carolyn. 1983. *The Invisible Farmers: Women in Agricultural Production*. Totowa, N.J.: Rowman and Allanheld.

———. 1996. *Gendered Fields: Rural Women, Agriculture, and Environment*. Boulder, Colo.: Westview Press.

Salisbury, Joyce. 1994. *The Beast Within: Animals in the Middle Ages*. New York: Routledge.

Sanders, Clinton. 1994. "Annoying Owners: Routine Interactions with Problematic Clients in a General Veterinary Practice." *Qualitative Sociology* 17 (2): 159–170.

———. 2006. "'The Dog You Deserve': Ambivalence in the K-9 Officer/Patrol Dog Relationship." *Journal of Contemporary Ethnography* 35 (2): 148–172.

Sandøe, Peter, Stine Christiansen, and Björn Forkman. 2006. "Animal Welfare: What Is the Role of Science?" In *Animals, Ethics and Trade: The Challenge of Animal Sentience*, ed. Jacky Turner and Joyce D'Silva, 41–52. London: Earthscan.

Saugeres, Lise. 2002a. "The Cultural Representation of the Farming Landscape: Masculinity, Power and Nature." *Journal of Rural Studies* 18 (4): 373–384.

———. 2002b. "Of Tractors and Men: Masculinity, Technology and Power in a French Farming Community." *Sociologia Ruralis* 42 (2): 143–159.

———. 2002c. "'She's Not Really a Woman, She's Half a Man': Gendered Discourses of Embodiment in a French Farming Community." *Women's Studies International Forum* 25 (6): 641–650.

Schlosser, Eric. 2002. *Fast Food Nation: What the All-American Meal Is Doing to the World*. London: Penguin Books.

Schwabe, Calvin. 1994. "Animals in the Ancient World." In *Animals and Human Society: Changing Perspectives*, ed. Aubrey Manning and James Serpell, 36–58. London: Routledge.

Schwarz, Ulrike. 2004. *"To Farm or Not to Farm": Gendered Paths to Succession and Inheritance*. Münster, Germany: Lit Verlag.

Scott, Marvin, and Stanford Lyman. 1968. "Accounts." *American Sociological Review* 33 (1): 46–62.

Seabrook, Martin. 1994. "The Effect of Production Systems on the Behaviour and Attitudes of Stockpersons." In *Biological Basis of Sustainable Animal Production*, ed. E. Huisman, J. Osse, D. van der Heide, S. Tamminga, B. Tolkamp, W. Schouten, C. Hollingworth, and G. van Winkel, 252–258. Wageningen, Netherlands: Wageningen Pers.

Serpell, James. 1989. "Pet-Keeping and Animal Domestication: A Reappraisal." In *The Walking Larder: Patterns of Domestication, Pastoralism, and Predation*, ed. Juliet Clutton-Brock, 10–21. London: Unwin Hyman.

———. 1996. *In the Company of Animals: A Study of Human–Animal Relationships*. Cambridge: Cambridge University Press.

———. 1998a. "Companion Animals and Pets." In *Encyclopedia of Animal Rights and Animal Welfare*, ed. Marc Bekoff, 111–112. Westport, Conn.: Greenwood Press.

———. 1998b. "Domestication." In *Encyclopedia of Animal Rights and Animal Welfare*, ed. Marc Bekoff, 136–138. Westport, Conn.: Greenwood Press.

———. 2003. "Anthropomorphism and Anthropomorphic Selection—beyond the 'Cute Response.'" *Society and Animals* 11 (1): 83–100.

———. 2004. "Factors Influencing Human Attitudes to Animals and their Welfare." *Animal Welfare* 13 (supp.): 145–151.

Seymour, John. 1996. *The Complete Book of Self-Sufficiency*. London: Dorling Kindersley.

Shapiro, Kenneth. 1989. "The Death of the Animal: Ontological Vulnerability." *Between the Species* 5:183–194.

———. 1994. "The Caring Sleuth: Portrait of an Animal Rights Activist." *Society and Animals* 2 (2): 145–165.

Shortall, Sally. 2000. "In and Out of the Milking Parlour: A Cross-National Comparison of Gender, the Dairy Industry and the State." *Women's Studies International Forum* 23 (2): 247–257.

———. 2001. "Women in the Field: Women, Farming and Organizations." *Gender, Work and Organization* 8 (2): 164–181.

———. 2002. "Gendered Agricultural and Rural Restructuring: A Case Study of Northern Ireland." *Sociologia Ruralis* 42 (2): 160–175.

Sinclair, Upton. 1906. *The Jungle*. London: Penguin Books.

Singer, Peter. 1995. *Animal Liberation*, 2nd ed. London: Pimlico.

Skeat, Walter. 1910. *An Etymological Dictionary of the English Language*. Oxford: Clarendon Press.

Skjelbred, A. 1994. "Milk and Milk Products in a Woman's World." In *Milk and Milk Products: From Medieval to Modern Times*, ed. P. Lysaght, 198–207. Dublin: Canongate Academic.

Slee, Bill. 1998. "Unwanted and Unloved." *Independent*, November 18, p. 5.

Smith, Allen C., III, and Sherryl Kleinman. 1989. "Managing Emotions in Medical School: Students' Contacts with the Living and the Dead." *Social Psychology Quarterly* 52 (1): 56–69.

Smith, Charles. 1989. *Auctions: The Social Construction of Value*. Berkeley: University of California Press.

Smith, Mick. 2002. "The 'Ethical' Space of the Abattoir: On the (In)humane Slaughter of Other Animals." *Human Ecology Review* 9 (2): 49–58.

Smith, Peter, and Ray Bradley. 2003. "Bovine Spongiform Encephalopathy (BSE) and Its Epidemiology." *British Medical Bulletin* 66 (1): 185–198.

Smith, Rebecca. 2000. *Sentenced to Death: A Viva! Report on the Slaughter of Farmed Animals in the U.K.* Brighton, U.K.: Viva!

Sommestad, Lena, and Sally McMurry. 1998. "Farm Daughters and Industrialization: A Comparative Analysis of Dairying in New York and Sweden, 1860–1920." *Journal of Women's History* 10 (2): 137–164.

Soul, Peter. 1998. "The Meat Hygiene Service: Protecting Food Safety and Animal Welfare." *Continuing Professional Development Veterinary Medicine* 1 (2): 64–67.

Statutory Instrument. 1995. *The Specified Bovine Offal Order 1995*. No. 1928. London: Her Majesty's Stationery Office.

———. 1998. *The Cattle Identification Regulations*. No. 871. London: Her Majesty's Stationery Office.

Steinfeld, Henning. 2004. "The Livestock Revolution—a Global Veterinary Mission." *Veterinary Parasitology* 125 (1–2): 19–41.

Stevenson, Peter. 1994. *A Far Cry from Noah: The Live Export Trade in Calves, Sheep and Pigs*. London: Merlin Press.

———. 1997. *Factory Farming and the Myth of Cheap Food: The Economic Implications of Intensive Animal Husbandry Systems*. Hampshire, U.K.: Compassion in World Farming Trust.

Stookey, Joseph, and Jon Watts. 2004. "Production Practices and Well-being: Beef Cattle." In *The Well-being of Farm Animals: Challenges and Solutions*, ed. John Benson and Bernard Rollin, 185–205. Oxford: Blackwell.

Swabe, Joanna. 1999. *Animals, Disease and Human Society: Human–Animal Relations and the Rise of Veterinary Medicine*. London: Routledge.

———. 2000. "Veterinary Dilemmas: Ambiguity and Ambivalence in Human–Animal Interaction." In *Companion Animals and Us: Exploring the Relationships between People and Pets*, ed. Anthony Podberscek, Elizabeth Paul, and James Serpell, 292–312. Cambridge: Cambridge University Press.

Sykes, Gresham, and David Matza. 1957. "Techniques of Neutralization: A Theory of Delinquency." *American Sociological Review* 22 (6): 664–670.

Symon, J. 1959. *Scottish Farming: Past and Present*. Edinburgh: Oliver and Boyd.

Tansey, Geoff, and Joyce D'Silva, eds. 1999. *The Meat Business: Devouring a Hungry Planet*. London: Earthscan.

Thomas, Keith. 1983 *Man and the Natural World: Changing Attitudes in England, 1500–1800*. London: Penguin Books.

Thompson, Paul. 2004. "Getting Pragmatic about Farm Animal Welfare." In *Animal Pragmatism: Re-thinking Human–Nonhuman Relationships*, ed. Erin Mckenna and Andrew Light, 140–159. Bloomington: Indiana University Press.

Thompson, William. 1983. "Hanging Tongues: A Sociological Encounter with the Assembly Line." *Qualitative Sociology* 6 (3): 215–237.

Thomsen, P. T, and J. T. Sørensen. 2009. "Factors Affecting the Risk of Euthanasia for Cows in Danish Dairy Herds." *Veterinary Record* 165 (2): 43–45.

Thomson, John. 2005. *Ring of Memories: Fairs and Livestock Auction Marts of Scotland.* Anan, Scotland: J. and M. Thomson.

Thomson, Steven. 2008. "Foot and Mouth Disease Review: Structure of the Scottish Livestock Industry." AA211 Special Study Report for the Scottish Government's Rural and Environment Research and Analysis Directorate, Scottish Agricultural College, Edinburgh. Available online at http://www.scotland.gov.uk (accessed December 20, 2009).

Todhunter, J. 1995. "Ignorance Is No Excuse." *Ark* (November): 418.

Tolson, Andrew. 1977. *The Limits of Masculinity*. London: Tavistock.

Toussaint-Samat, Maguelonne. 1994. *A History of Food*. Cambridge: Blackwell.

Tovey, Hilary. 2003. "Theorising Nature and Society in Sociology: The Invisibility of Animals." *Sociologia Ruralis* 43 (3): 196–215.

Trauger, Amy. 2001. "Women Farmers in Minnesota and the Post-productivist Transition." *Great Lakes Geographer* 8 (2): 53–66.

———. 2004. "'Because They Can Do the Work': Women Farmers in Sustainable Agriculture in Pennsylvania, USA." *Gender, Place and Culture* 11 (2): 289–307.

Trow-Smith, Robert. 1959. *A History of British Livestock Husbandry 1700–1900*. London: Routledge and Kegan Paul.

Tuan, Yi-fu. 1984. *Dominance and Affection: The Making of Pets*. New Haven, Conn.: Yale University Press.

Turner, Jacky. 2003. *Stop—Look—Listen: Recognising the Sentience of Farm Animals.* Hampshire, U.K.: Compassion in World Farming Trust.

Turner, Jacky, and Joyce D'Silva. 2006. *Animals, Ethics and Trade: The Challenge of Animal Sentience.* London: Earthscan.

TV Comedy Resources. 2008. "British TV Comedy—*The Good Life*." Available online at http://www.phill.co.uk/comedy/goodlife (accessed September 20, 2008).

Twine, Richard. 2007. "Animal Genomics and Ambivalence: A Sociology of Animal Bodies in Agricultural Biotechnology." *Genomics, Society and Policy* 3 (2): 99–117.

Union of Concerned Scientists. 2006. "European Union Bans Antibiotics for Growth Promotion." Available online at http://www.ucsusa.org (accessed March 21, 2009).

Vialles, Noelie. 1994. *Animal to Edible*. Cambridge: Cambridge University Press.

Waiblinger, Susanne, Xavier Boivin, Vivi Pedersen, Maria-Vittoria Tosi, Andrew Janczak, E. Kathalijne Visser, and Robert Jones. 2006. "Assessing the Human–Animal Relationship in Farmed Species: A Critical Review." *Applied Animal Behaviour Science* 101 (3–4): 185–242.

Walbert, David. 1997–2004. "The New Agrarian. Hobby Farming." Available online at http://www.newagrarian.com (accessed September 20, 2008).

Walker, B. 1986. "The Flesher's Trade in Eighteenth- and Nineteenth-Century Scotland: An Exploratory Investigation into Slaughtering Techniques, Tools and Buildings." In *Food in Change: Eating Habits from the Middle Ages to the Present Day*, ed. Alexander Fenton and Eszter Kisbán, 127–137. Edinburgh: John Donald Publishers.

Wallace, Claire, Pamela Abbott, and Gloria Lankshear. 1996. "Women Farmers in South-West England." *Journal of Gender Studies* 5 (1): 49–62.

Walton, John. 1984. "The Rise of Agricultural Auctioneering in Eighteenth- and Nineteenth-Century Britain." *Journal of Historical Geography* 10 (1): 15–36.

———. 1999. "Pedigree and Productivity in the British and North American Cattle Kingdoms before 1930." *Journal of Historical Geography* 25 (4): 441–462.

Warrick, Joby. 2001. "They Die Piece by Piece: In Overtaxed Plants, Humane Treatment of Cattle Is Often a Battle Lost." *Washington Post*, April 10.

Wathes, C. 1994. "Animals in Man's Environment: A Question of Interest." *Outlook on Agriculture* 23 (1): 47–54.

Watts, Mark. 1999. "An Agenda for Reform: Farm Animal Welfare in the European Union." In *The Meat Business: Devouring a Hungry Planet*, ed. Geoff Tansey and Joyce D'Silva, 65–72. London: Earthscan.

Webster, John. 1994. *Animal Welfare: A Cool Eye towards Eden*. Oxford: Blackwell Science.

———. 2005. *Animal Welfare: Limping towards Eden*. Oxford: Blackwell.

Weigert, Andrew. 1991. *Mixed Emotions: Certain Steps toward Understanding Ambivalence*. Albany: State University of New York Press.

Wharton, Amy S. 2009. "The Sociology of Emotional Labor." *Annual Review of Sociology* 35:147–65.

Whatmore, Sarah. 1991. *Farming Women: Gender, Work and Family Enterprise*. Hampshire, U.K.: Macmillan Academic and Professional.

Wilkie, Rhoda. 2002. "Sentient Commodities: Human–Livestock Relations from Birth to Slaughter in Commercial and Hobby Production." Ph.D diss., Aberdeen University.

———. 2005. "Sentient Commodities and Productive Paradoxes: The Ambiguous Nature of Human–Livestock Relations in Northeast Scotland." *Journal of Rural Studies* 21 (2): 213–230.

———. 2006. "Sentient Commodities: A Sociological Exploration of Human–Food Animal Interactions." *Gibson Institute for Land, Food and Environment, Research Paper Series* 2 (3): 1–40.

Williams, Anna. 2004. "Disciplining Animals: Sentience, Production, and Critique." *International Journal of Sociology and Social Policy* 24 (9): 45–57.

Williams, E. 1976. "The Development of the Meat Industry." In *The Making of the Modern British Diet*, ed. Derek Oddy and Derek. Miller, 44–57. London: Croom Helm.

Wilson, Geoff. 2001. "From Productivism to Post-productivism . . . and Back Again? Exploring the (Un)changed Natural and Mental Landscapes of European Agriculture." *Transactions of the Institute of British Geographers* 26 (1): 77–102.

———. 2008. "From 'Weak' to 'Strong' Multifunctionality: Conceptualising Farm-Level Multifunctional Transitional Pathways." *Journal of Rural Studies* 24 (3): 367–383.

Wilson, Peter. 2007. "Agriculture or Architecture? The Beginnings of Domestication." In *Where the Wild Things Are Now: Domestication Reconsidered*, ed. Rebecca Cassidy and Molly Mullin, 101–121. Oxford: Berg.

Winders, Bill, and David Nibert. 2004. "Consuming the Surplus: Expanding 'Meat' Consumption and Animal Oppression." *International Journal of Sociology and Social Policy* 24 (9): 76–96.

Winter, Michael, C. Fry, and S. Carruthers. 1998. "European Agricultural Policy and Farm Animal Welfare." *Food Policy* 23 (3–4): 305–323.

Wiskerke, J. 2003. "On Promising Niches and Constraining Sociotechnical Regimes: The Case of Dutch Wheat and Bread." *Environment and Planning* A 35 (3): 429–448.

Wolfson, David, and Mariann Sullivan. 2004. "Foxes in the Hen House: Animals, Agribusiness, and the Law." In *Animal Rights: Current Debates and New Directions,* ed. Cass Sunstein and Martha Nussbaum, 205–233. Oxford: Oxford University Press.

Woods, Abigail. 2004. "Why Slaughter? The Cultural Dimensions of Britain's Foot and Mouth Disease Control Policy, 1892–2001." *Journal of Agricultural and Environmental Ethics* 17 (4–5): 341–362.

———. 2007. "The Farm as Clinic: Veterinary Expertise and the Transformation of Dairy Farming, 1930–1950." *Studies in History and Philosophy of Biological and Biomedical Sciences* 38 (2): 462–487.

Wooldridge, G. 1922. "The Humane Slaughtering of Animals." *Veterinary Record* 2 (9): 139–147.

World Health Organization. 2002. "Bovine Spongiform Encephalopathy." Fact Sheet no. 113. Available online at http://www.who.int (accessed August 23, 2009).

Wright, Jim, Tim Stephens, Ron Wilson, and Julian Smith. 2002. "The Effect of Local Livestock Population Changes on Auction Market Viability—a Spatial Analysis." *Journal of Rural Studies* 18 (4): 477–483.

Yancy, Allen. 2006. "Veterinarians and the Case against Legal Personhood for Animals." In *People Property or Pets?* ed. Marc Hauser, Fiery Cushman, and Matthew Kamen, 197–204. West Lafayette, Ind.: Purdue University Press.

Yarwood, Richard, and Nick Evans. 1998. "New Places for 'Old Spots': The Changing Geographies of Domestic Livestock Animals." *Society and Animals* 6 (2): 137–165.

———. 1999. "The Changing Geography of Rare Livestock Breeds in Britain." *Geography: This Changing World* 84 (1): 80–91.

———. 2000. "Taking Stock of Farm Animals and Rurality." In *Animal Spaces, Beastly Places: New Geographies of Human–Animal Relations,* ed. Chris Philo and Chris Wilbert, 98–114. London: Routledge.

———. 2006. "A Lleyn Sweep for Local Sheep? Breed Societies and the Geographies of Welsh Livestock." *Environment and Planning* A 38 (7): 1307–1326.

Yarwood, Richard, Nick Evans, and Julie Higginbottom. 1997. "The Contemporary Geography of Indigenous Irish Livestock." *Irish Geography* 30 (1): 17–30.

Zeuner, Frederick. 1963a. *A History of Domesticated Animals.* London: Hutchinson of London.

———. 1963b. "The History of the Domestication of Cattle." In *Man and Cattle: Proceedings of a Symposium on Domestication at the Royal Anthropological Institute,* ed. Frederick Zeuner and Arthur Mourant, 9–19. London: Royal Anthropological Institute.

Index

The letter t *following a page number denotes a table.*

Rhoda M. Wilkie is a lecturer in sociology at the University of Aberdeen, where she earned her doctorate in 2002. She is the coeditor (with David Inglis) of the five-volume collection *Animals and Society: Critical Concepts in the Social Sciences.*